Irrigation Systems and Applications

Edited by Muhammad Sultan, Muhammad Imran and Fiaz Ahmad

Published in London, United Kingdom

Irrigation Systems and Applications
http://dx.doi.org/10.5772/intechopen.107802
Edited by Muhammad Sultan, Muhammad Imran and Fiaz Ahmad

Contributors
Abel Ogunleye, Aliev Nozim Numonovich, Aneela Sarwar, Arulanandam Mariadoss, Ashutosh Sawarkar, B.Sahadeva Reddy, Bishal Mukherjee, C. Nagamani, Elangovan Subramanian, Fiaz Ahmad, Hadeed Ashraf, Hafiz M. Asfahan, K.Madhusudhan Reddy, Kayode Adejobi, Michael Thomas Aide, Muhammad Adnan Shahid, Muhammad Aleem, Muhammad Danish Majeed, Muhammad Imran, Muhammad Safdar, Muhammad Sultan, Narayanaswamy Nithya, Pasoubady Saravanane, Pavan Kumar Reddy Yerasi, Rehan Mehmood Sabir, Samuel Agele, Shanmugam Vijayakumar, Suman Dutta, Syeda Mishal Zahra, V.Siva Jyothi, Zafar A. Khan, Zahid Maqbool

First published in London, United Kingdom, 2024 by IntechOpen
IntechOpen is the global imprint of INTECHOPEN LIMITED, registered in England and Wales, registration number: 11086078, 167-169 Great Portland Street, London, W1W 5PF, United Kingdom

British Library Cataloguing-in-Publication Data
A catalogue record for this book is available from the British Library

Additional hard and PDF copies can be obtained from orders@intechopen.com

Irrigation Systems and Applications
Edited by Muhammad Sultan, Muhammad Imran and Fiaz Ahmad
p. cm.

This title is part of the Agricultural Sciences Book Series, Volume 6
Topic: Agronomy and Horticulture
Series Editor: W. James Grichar
Topic Editor: Ibrahim Kahramanoglu
Associate Topic Editors: Murat Helvaci and Olga Panfilova

Print ISBN 978-1-83769-057-2
Online ISBN 978-1-83769-058-9
eBook (PDF) ISBN 978-1-83769-059-6
ISSN 3029-052X

For EU product safety concerns:
IN TECH d.o.o., Prolaz Marije Krucifikse Kozulić 3, 51000 Rijeka, Croatia,
info@intechopen.com or visit our website at intechopen.com.

IntechOpen Book Series
Agricultural Sciences
Volume 6

Aims and Scope of the Series

The importance of agriculture cannot be overstated. It helps sustain life, as it gives us the food we need to survive and provides opportunities for economic well-being. Agriculture helps people prosper around the world and combines the creativity, imagination, and skill involved in planting crops and raising animals with modern production methods and new technologies. This series includes two main topics: Agronomy and Horticulture, and Animal Farming. This series will help readers better understand the intricacies of production agriculture and provide the new knowledge that is required to be successful. The success of a farmer in modern agriculture requires knowledge of events happening locally as well as globally that impact input decisions and ultimately determine net profit.

Meet the Series Editor

W. James Grichar has been employed with Texas A&M AgriLife Research for over 45 years with an emphasis on research in agronomy, plant pathology, and weed science. He obtained his BS from Texas A&M in 1972 and his Masters of Plant Protection in 1975. He has published 195 journal articles, over 330 research reports and briefs, 11 book chapters, and over 300 abstracts of profession meetings. He also directs research in many crops including corn, grain sorghum, peanuts, and sesame. He has held various positions in different professional societies including the American Peanut Research and Education Society, Southern Weed Science Society, and Texas Plant Protection Conference in addition to being Associate Editor for Peanut Science and Weed Technology. Significant accomplishments have included spearheading efforts to determine the optimum planting time for soybean production along the upper Texas Gulf Coast. These efforts have shown growers that soybean yields can be improved by 10 to 20% by following a late March to early April plant date. He also has been instrumental in developing a herbicide program for peanut production in the south Texas growing region. Through the development and use of herbicides that are effective against major weed problems in the south Texas region, peanut yields have increased by 25 to 30%.

Meet the Volume Editors

Dr. Muhammad Sultan is an accomplished Associate Professor of Agricultural Engineering at Bahauddin Zakariya University, Pakistan. Holding a BSc and MSc in Agricultural Engineering with honors from the University of Agriculture Faisalabad, Pakistan, he earned his Ph.D. from Kyushu University, Japan as a recipient of an MEXT fellowship from the Government of Japan. He also conducted postdoctoral research at Kyushu University, Japan, and Simon Fraser University, Canada as a recipient of Japan Student Services Organization (JASSO) and Canadian Queen Elizabeth fellowships. Recognized among the world's top 2% Scientists by Stanford University, USA and Elsevier, Dr. Sultan is the leading scientist in Pakistan for agricultural engineering, according to the AD Scientific Index Ranking. With a vast publication record, he serves as editor for various high-impact journals, contributing significantly to the field of agricultural engineering. Engaged in multiple research projects and revered for supervising theses, Dr. Sultan continues to shape the landscape of energy-efficient agricultural technologies.

Dr. Muhammad Imran is an associate professor at Aston University, UK, and a distinguished researcher renowned for his expertise in sustainable thermal energy systems, specializing in heating, cooling, and distributed energy applications. Dr. Imran's recent leadership in two innovative projects focused on novel and smart irrigation systems powered by hybrid energy solutions showcases a commitment to advancing sustainable practices in agriculture. As the principal investigator for projects in Ethiopia and Kenya, Dr. Imran is spearheading efforts to integrate cutting-edge cooling technologies into agricultural practices, addressing post-harvest losses in developing countries. Beyond research, Dr. Imran holds leadership roles at Aston University, actively engages with industry, and contributes to the academic community through publications and supervisory roles.

Dr. Fiaz Ahmad is an associate professor in the Department of Agricultural Engineering, Bahauddin Zakariya University, Multan, Pakistan. He obtained his Ph.D. in Agricultural Bioenvironmental and Energy Engineering from Nanjing Agriculture University, China, in 2015, and a postdoctorate degree in Agricultural Engineering from Jiangsu University, Zhenjiang, China, in 2020. He earned a BSc and MSc in Agricultural Engineering from the University of Agriculture, Faisalabad, Pakistan, in 2001 and 2007, respectively. Dr. Ahmad is the author of more than seventy journal and conference articles. He has supervised more than ten master's students to date. In addition, he is currently supervising five master's and two doctoral students. He completed five research projects and is working on three others. His research interests include various aspects of agricultural and farm mechanization.

Contents

Preface

We are delighted to present this edited volume, *Irrigation Systems and Applications*, which brings together a diverse array of perspectives on the crucial intersection of irrigation technology and agricultural practices. As the editors of this comprehensive work, we find great pleasure in introducing the readers to a wealth of knowledge encapsulated in the following eight chapters, each contributing significantly to our understanding of irrigation systems and their wide-ranging applications.

This edited book is organized into two distinct sections, each containing four chapters that collectively provide a holistic view of the current state and future prospects of irrigation in agriculture. Section I, "Irrigation Systems", opens with Chapter 1, exploring the implementation of a hybrid energy-powered smart irrigation system tailored for smallholder farmers. The subsequent chapters delve into the evolution of sustainable rice production (Chapter 2), the utilization of hydroponics for enhanced agricultural productivity (Chapter 3), and the intricacies of advanced micro irrigation techniques (Chapter 4).

Transitioning to Section II, "Irrigation Applications", Chapter 5 investigates the monitoring of irrigated lands in the Hissar Valley of the Republic of Tajikistan. Chapter 6 explores the role of smallholder irrigation in climate mitigation and the improvement of cacao performance in the rainforest tropics. The volume continues with Chapter 7, which focuses on revolutionizing rice farming to maximize yield with minimal water, addressing the imperative of sustaining a hungry planet. Finally, Chapter 8 delves into the application of geospatial techniques in agricultural resource management.

This collection serves as a vital resource for researchers, students, and professionals seeking a nuanced understanding of the challenges and opportunities associated with irrigation systems and their applications. The contributors bring forth their expertise to illuminate various facets of smart farming, sustainable practices, and innovative technologies that have the potential to reshape agricultural landscapes globally.

As the editors, we express our gratitude to the esteemed authors who have shared their insights, making this edited book a valuable contribution to the field. We believe that the diverse perspectives presented in these chapters will inspire further research and innovation in the domain of irrigation, ultimately contributing to the sustainable growth of agricultural practices worldwide.

We hope this volume sparks curiosity and fosters a deeper appreciation for the intricate relationship between irrigation systems and their applications, setting the stage for future advancements in agricultural practices.

Dr. Muhammad Sultan and Dr. Fiaz Ahmad
Bahauddin Zakariya University,
Multan, Pakistan

Dr. Muhammad Imran
Aston University,
Birmingham, United Kingdom

Section 1

Irrigation Systems

Chapter 1

Hybrid Energy Powered Smart Irrigation System for Smallholder Farmers: Installation Site and Crop Selection

Muhammad Aleem, Muhammad Sultan, Muhammad Imran, Zafar A. Khan, Hadeed Ashraf, Hafiz M. Asfahan and Fiaz Ahmad

Abstract

In the context of food-energy-water nexus and uncertainties in climate change, hybrid energy powered smart irrigation system (HEPSIS) is an emerging solution for optimizing both energy and water to boost crop yield. In Pakistan, most of the farmers especially smallholder farmers are currently relying on conventional irrigation practices which result in high water consumptions, high energy consumptions (by means of pumping), low crop yields, and net profit. Prior to design/development, installation, and testing of the HEPSIS, it is essential to know a suitable site and potential food/cash crops which will be irrigated. In this regard, the study aims to select installation site and potential crops. Site suitability is explored for Sindh province from viewpoints of Indus Basin Irrigation System mapping, groundwater table depth/quality mapping, land use land cover, and soil classifications. Furthermore, crop selection analyses are performed by means of a screening matrix approach based on stars to identify two potential food and cash crops. As per the results, Badin, Ghotki, Khairpur, Sanghar, Shikarpur, Larkana, and Thatta are selected as some suitable sites for the proposed HEPSIS. Additionally, wheat and rice are selected as potential food crops whereas cotton and sugarcane are selected as potential cash crops.

Keywords: food-energy-water nexus, smart irrigation, site suitability, crops selection, food/cash crops, Sindh

1. Introduction

1.1 Background

The agriculture sector plays an essential role in the economy of Pakistan as it contributes about 22.7% of the gross domestic product (GDP) as well as food security. However, the contribution of the agriculture sector to the country's GDP has been steadily decreasing from 30.4% to 22.7% since 1980 to 2022 [1] because of climate-induced desertification. On the other hand, water scarcity and food demand are increasing corresponding to an increase in the population of the country [2–4]. Currently, Pakistan is ranked 12th among the most vulnerable countries where agriculture and water resource sectors are significantly influenced by climate change (CC) and it is projected that underdeveloped countries like Pakistan will experience more severe effects on these sectors in upcoming decades [5–7]. In these perspectives, climate-smart adaptation strategies are necessary for sustainability in the agriculture sector in terms of food security [8]. **Figure 1** shows CC's impact on the agriculture sector and climate-smart adaptation strategies. Among various subsectors of agriculture, the crop sector is an important subsector that contributes about 4.41% to the GDP of the country and ensures food security for a rapidly growing population [1, 10]. The

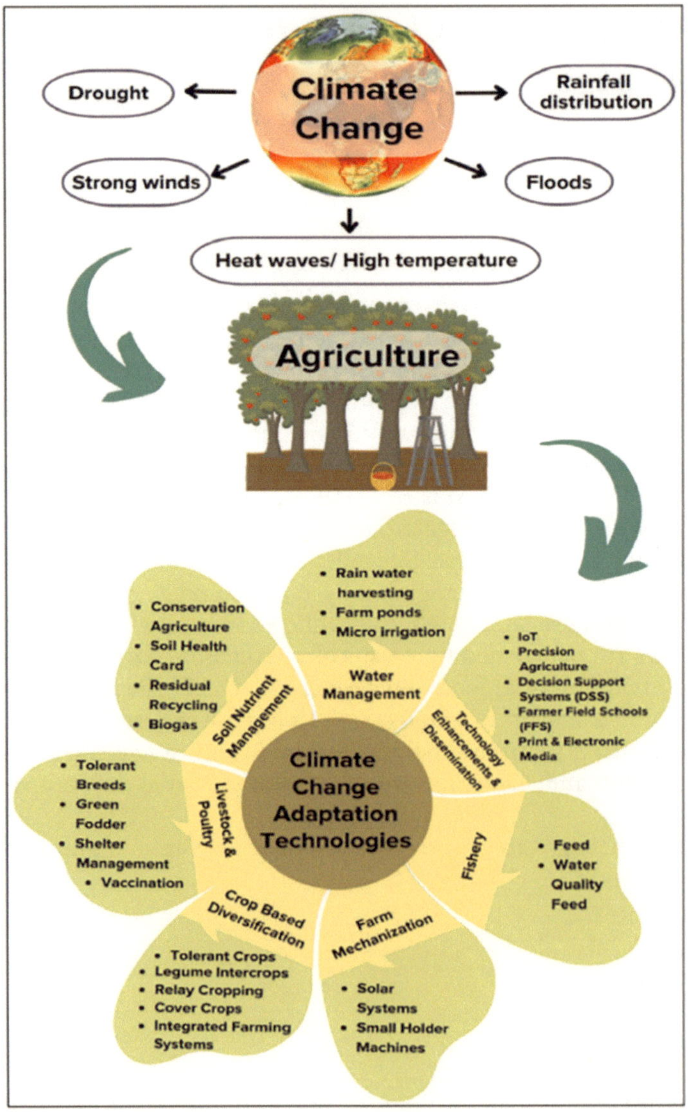

Figure 1.
Climate change impacts on agriculture sector with associated climate-smart adaptation strategies [9].

country's crop sector is primarily dependent on irrigation i.e., artificial application of water to the soil to meet crop water requirements (CWR) that vary with geography, types of crops, and their growing stages [11–13]. About 90% of freshwater is used by irrigation practices for meeting the CWR. This excessive amount of freshwater consumption exerts huge pressure on freshwater reserves. Additionally, irrigation practices consume a significant portion of energy by means of pumping which results in the emission of greenhouse gases [14]. **Figure 2** shows the volume of water utilized for irrigation via pumping, amount of energy consumed for pumping irrigation water, irrigation energy footprint, and irrigation carbon footprint for major crops in Pakistan. Therefore, a trade-off between energy utilization and supply of irrigation water is essential for sustainable agriculture.

Indus Basin Irrigation System (IBIS) is the largest component and source of irrigation. However, irrigation system in Pakistan is facing major challenges like depleting water reserves, canal water losses, inappropriate water management practices, and lack of access to modern techniques, thereby reducing crop yields [7, 15, 16]. Different kinds of irrigation systems like sprinklers and drips are being employed worldwide. However, farmers across a major portion of the country employed warabandi system (using canal water) to apply irrigation without knowing the actual CWR. Furthermore, conventional irrigation systems are constrained by salinity, water logging, and low application efficiency [17, 18]. In this regard, renewable energy-operated smart irrigation systems can address the aforesaid dilemma [19]. The smart irrigation system

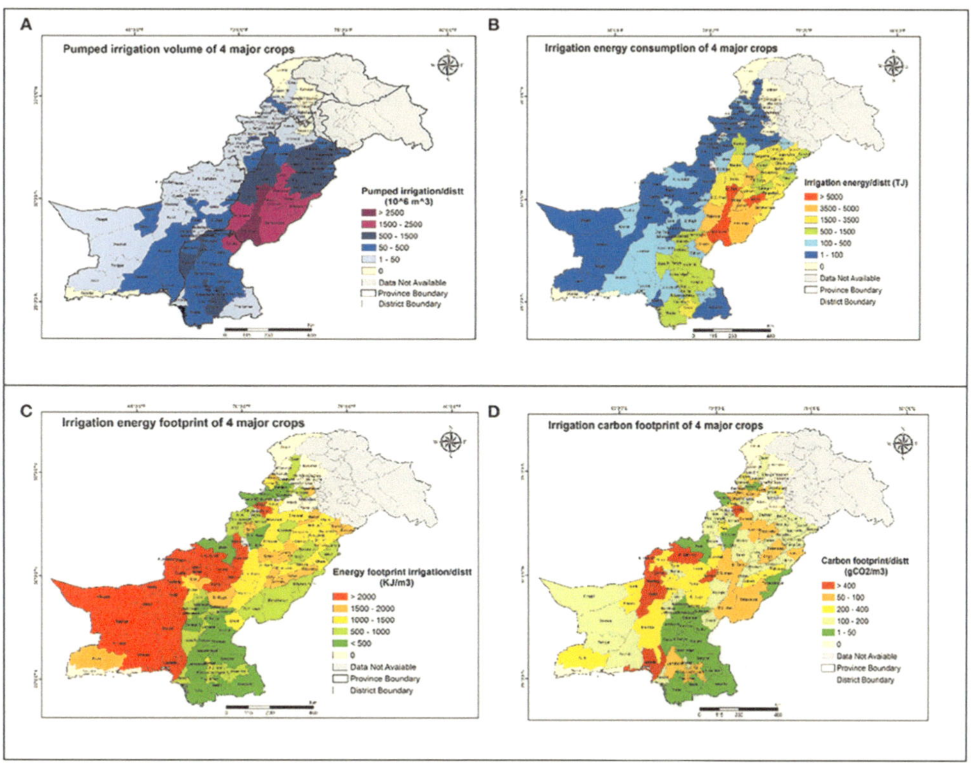

Figure 2.
Pakistan's map shows (a) volume of water utilized for irrigation via pumping, (b) amount of energy consumed for pumping irrigation water, (c) irrigation energy footprint, and (d) irrigation carbon footprint for major crops, reproduced here from [14].

can apply water based on the actual CWR corresponds to growth stages, thereby optimizing water utilization and enhancing crop yield.

1.2 Hybrid energy powered smart irrigation system

Hybrid energy powered smart irrigation system (HEPSIS) involves integration of renewable energy sources (solar/wind), sensor technology, artificial intelligence (AI), machine learning (ML), and internet of thing (IoT) based decision support systems (DSS) as well as mobile app (act as user interface) for controlling and monitoring mechanisms remotely in order to optimize energy and water use in irrigation [20–22]. **Figure 3** shows a working scheme of the HEPSIS. The solar/winds energy ensures a consistent and environment-friendly power supply which can be utilized to operate controllers and pumps for the distribution of irrigation water. The sensors help in collecting real-time data of climatic parameters including temperature, relative humidity, precipitation, wind speed, sunshine hours, solar intensity, and soil moisture levels which depend on the CWR [23]. The AI, ML, and IoT-based DSS receives collected data from the sensors and performs data processing to make decision about irrigation scheduling (when and how much to irrigate) [22, 24]. **Figure 4** provides a conceptual scheme of IoT in the smart irrigation system. In the end, the mobile app provides user-friendly interface to farmers remotely for monitoring and controlling irrigation practices and to make decisions accordingly.

The remote accessibility saves time and resources and overcomes the need for manual labour by increasing operational efficiency. Furthermore, integration of the HEPSIS with weather forecasting systems helps in boosting the effectiveness. By incorporating real-time climatic and soil moisture data, the HEPSIS can adapt

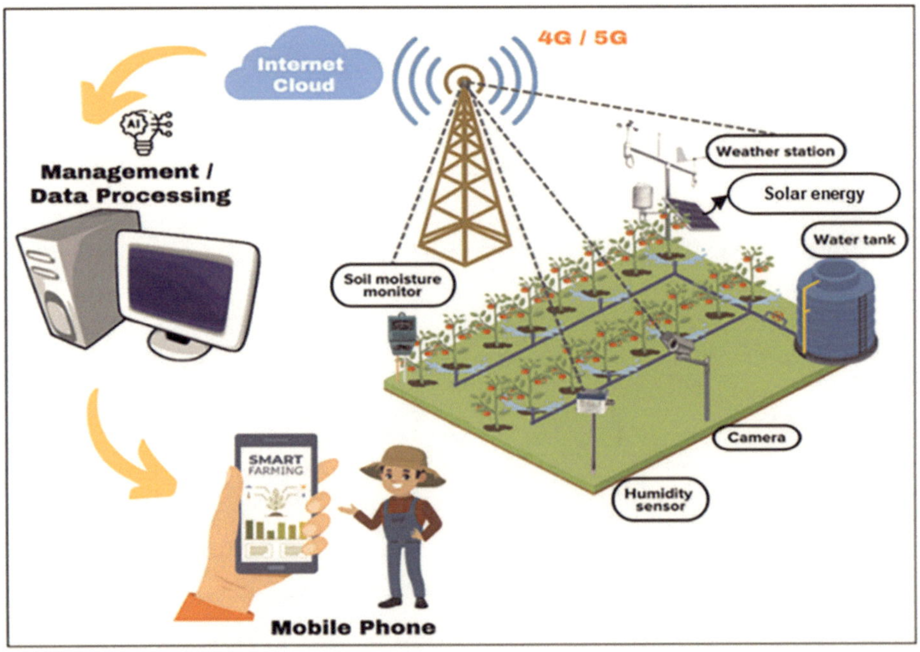

Figure 3.
Working scheme of HEPSIS, reproduced here from [9].

DOI: http://dx.doi.org/10.5772/intechopen.114144

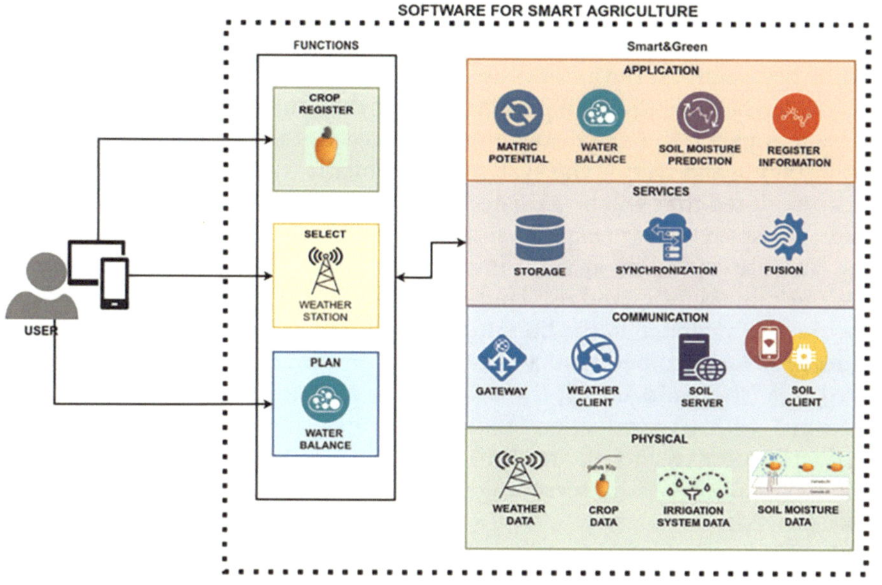

Figure 4.
Conceptual scheme of internet of things (IoT) for the smart irrigation system [25].

irrigation scheduling based on predicted temperature, and precipitation changes, thereby optimizing water utilization and reducing dependency on manual adjustments [26, 27]. Some benefits of the smart irrigation system for rice crops over conventional systems are presented in **Figure 5**. By minimizing water usage, farmers

Issues of conventional rice cultivation

- Low water use efficiency (WUE)
- Emission of greenhouse gases
- Declining groundwater table
- Virtual water trade deficit
- Low nutrient and energy use efficiency

Water-saving methods

Direct seeded rice (DSR)	⟷	Saturated soil culture (SSC)
Alternate wetting and drying (AWD)	⟷	System of rice intensification (SRI)
Aerobic rice	⟷	Smart irrigation
Drip irrigated rice	⟷	Best management practices (Seed treatments, weed control, etc.)

	AWD	SSC	DSR	Aerobic rice	SRI	Drip	Smart Irrigation	CF
Water saving (%)	~ 25	25-30	~ 15	20-30	20-40	>50	30	-
Percolation	Low	Very low	Low	Low	Low	Very low	Low	More
Weed	More	More	More	More	More	Low	More	Low
Suitable ecosystem	I	I	R	I	I	I	I	I/R
CH_4 emission	Low	Low	Low	Low	Low	Low	Low	More
N_2O emission	More	More	More	More	More	Low	More	Low
GWP	Low	Low	Low	Low	Low	Very low	Low	More
Irrigation	WT <15 cm in tube	Shallow irrigation	Normal irrigation	Maintain SM in field capacity	Hair like crack	2-days interval	Based on SM	4 to 7 days interval
Irrigation depth (cm)	<5	3	<5	<5	<5	No flooding	<5	>10
Suitable soil	Loamy	Loamy	All	Loam to clay	All	Sandy	All	Clay
Labour requirement	Low	Low	Low	Low	More	Very low	Very low	More
Energy input	Low	Low	Low	Low	Low	Low	Low	More
Cost	Low	Low	Low	Low	Low	More	More	More

GWP – Global warming potential; CF – Conventional flooding; I – Irrigated; R – Rainfed; WT – Water Table; SM –Soil moisture

Figure 5.
Comparison between smart irrigation systems and conventional irrigation systems [28].

can reduce their overall operational costs, improve energy efficiency, and mitigate environmental impacts, such as water pollution and soil erosion.

In these perspectives, the authors are planning to design, develop, and test HEPSIS for smallholder farmers in Sindh province under the project number: 10039507 funded by Innovate UK. For execution of the project, it is essential to explore suitable sites for installing and testing the system. Additionally, potential food and cash crops need to be explored that will be irrigated with the proposed smart irrigation system. Therefore, the study aims to explore suitable installation sites and crop selection analyses. The site suitability analysis is explored from viewpoints of IBIS mapping, groundwater table/depth, land use land cover (LULC) classification, and soil classification which are available in the literature [29–32]. For crops selection, seven significant factors for each studied food and cash crop are considered which are listed as (i) cultivated area (thousand acres), (ii) irrigation requirements (mm/season), (iii) water cost (Rs/acre), (iv) net production (thousand tonnes), (v) average yield (kg/acre), (vi) production cost (Rs/acre), and (vii) net profit (Rs/acre). In order to select two potential food and cash crops screening matrix i.e., qualitative methodology based on stars is employed.

2. Materials and methods

Figure 6 shows a flow chart presenting a summary of the research methodology executed in this study. The methodology comprises of two processes namely: site suitability and crops selection. The extensive details of each process are given in the following subsections.

2.1 Study area

In this study, Sindh province is selected as a study area, ranked second in agricultural production in Pakistan. Among a total area of 140, 914 km^2, about 51.9% of the province is characterized as a rural area and 48.1% as an urban area.

A total number of districts is thirty as shown in **Figure** 7. According to Köppen climate classification, Sindh experiences a subtropical desert climate, thereby evaporation is higher as compared to other provinces in Pakistan. Mean annual rainfall varied between 100 to 200 mm specifically in the Lower Indus Plain [29]. It is located at an elevation of 24.59 m (80.68 ft) above sea level and in the southern part of the IBIS. The IBIS network enters Sindh at Guddu barrage and four canals namely: Begari, Desert pat feeder, Ghotki, and Rainee offtake. The Guddu barrage meets the agricultural water or irrigation demands of Jacobabad, Kashmore, Qamber Shahdadkot, Shikarpur, and Sukkur districts of Sindh and the Naseerabad district of Balochistan through its four canals.

2.2 Site suitability analysis

Referring to **Figure 6a**, site suitability analysis involves four steps namely: (1) IBIS mapping, (2) groundwater depth and quality, (3) land use land cover classification (LULC), and (4) soil classification. Each step for Sindh province was performed in literature. Therefore, the present study utilizes data of the IBIS mapping [29], groundwater depth [30], groundwater quality [30], LULC classification [31], and soil classification [32] from the cited literature for exploring suitable installation sites.

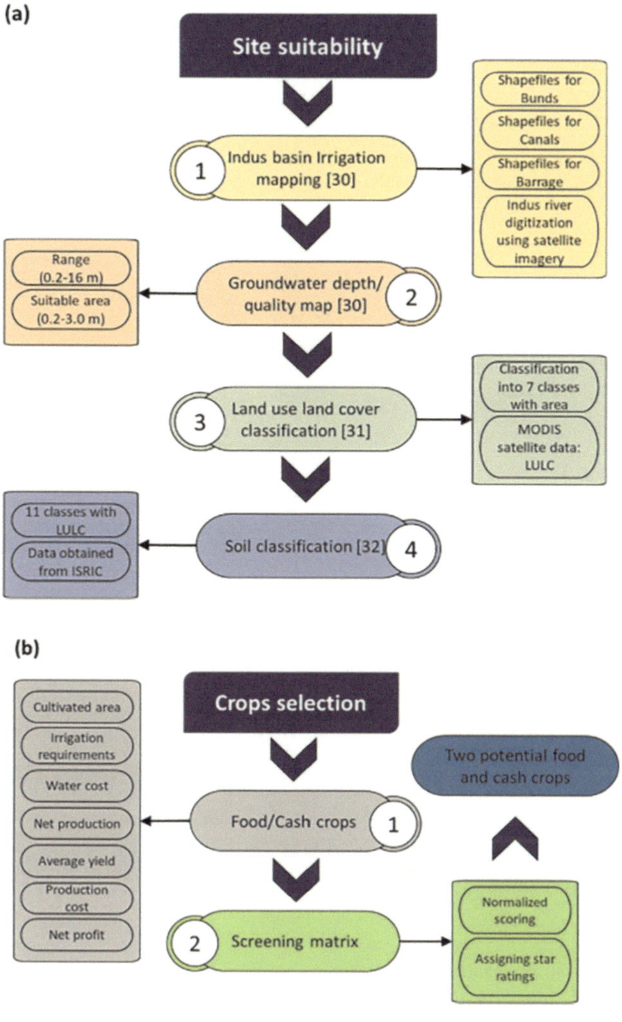

Figure 6.
A flow chart representing the summary of research methodology executed in this study (a) site suitability, and (b) crops selection.

2.3 Crops selection

Referring to **Figure 6b**, crops selection involves two steps namely: (1) data collection, and (2) screening matrix. Eight kinds of food and cash crops are selected in order to identify two potential food and cash crops. The food crops include wheat, rice, maize, moong, jowar, gram, onion, and tomato while the cash crops include cotton, sugarcane, rapeseed, chilies, sesame, potato, mango, and potato. Data of seven significant factors that are listed as (i) cultivated area (thousand acres), (ii) irrigation requirements (mm/season), (iii) water cost (Rs/acre), (iv) net production (thousand tonnes), (v) average yield (kg/acre), (vi) production cost (Rs/acre), and (vii) net profit (Rs/acre) are collected for each studied food and cash crop at national and provincial level specifically for Sindh. The data of each significant parameter excluding irrigation requirements for

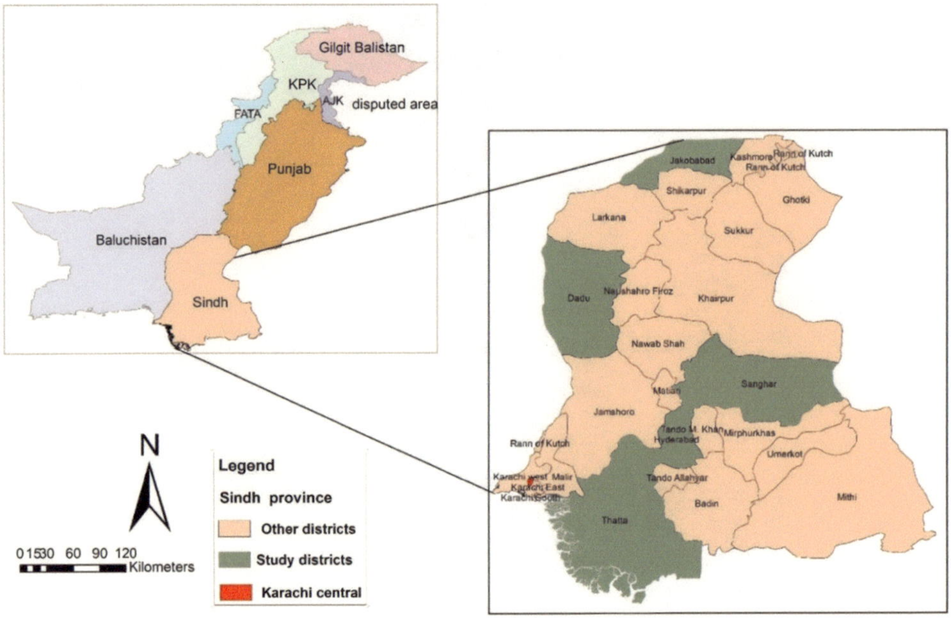

Figure 7.
Location and districts of the study area [33].

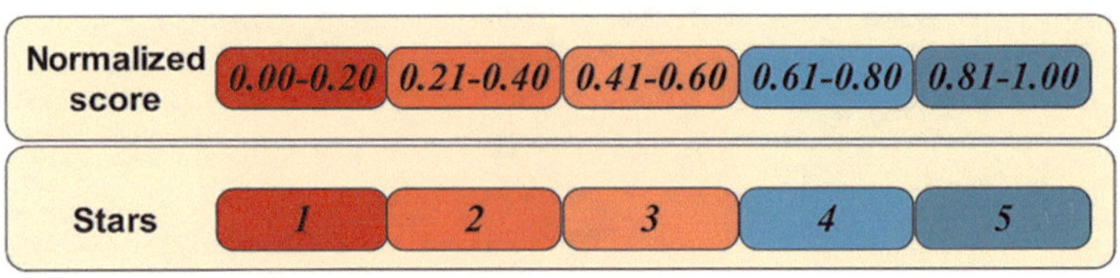

Figure 8.
Normalized score range and corresponding stars rating.

the studied crops is collected from the agricultural marketing information service [34]. The irrigation requirements data for each food and cash crop is obtained from food and agriculture organization (FAO).

Screening matrix methodology is a systematic approach used to estimate and compare various options based on specific criteria. The screening matrix involves assigning star ratings to each parameter in order to compute its suitability in accordance with the specified criteria. The star ratings serve to evaluate how well each option aligns with the defined criteria. A higher star rating indicates a stronger performance or better alignment with the desired criteria. In the context of selecting two potential food and cash crops, stars are assigned to each selected crop based on the normalized score obtained by the relative weightage of the significant factors. The weightage to each significant factor of selected crops is provided using a normalized scoring method as given by Eq. (1). The star ratings corresponding to normalized scores are provided in **Figure 8**.

$$Normalizedscore = \frac{(OriginalValue - Minimumvalue)}{(Maxiumvalue - Minimumvalue)} \tag{1}$$

3. Results and discussion

3.1 Site suitability analysis

3.1.1 IBIS

IBIS is considered as largest irrigation system throughout the globe, covering a total area of 17.2 million hectares with a length of 2900 km [35]. According to the World Bank Report, the Indus River basin is inhabited by over 300 million people across Afghanistan, China, India, and Pakistan. About half of the basin lies in Pakistan which covers approximately two-thirds of the land and 87% of the country's population [36]. The IBIS plays a significant role in meeting country's irrigation requirements and up to 96% of country's renewable water resources [36]. The river flows through the Himalayas at 18,000 ft to the Sindh plains and ultimately falls into the Arabian Sea. The basin experiences a mean annual flow of 176 billion m^3. About 90% of the flow is utilized for meeting irrigation requirements via canals passing through three

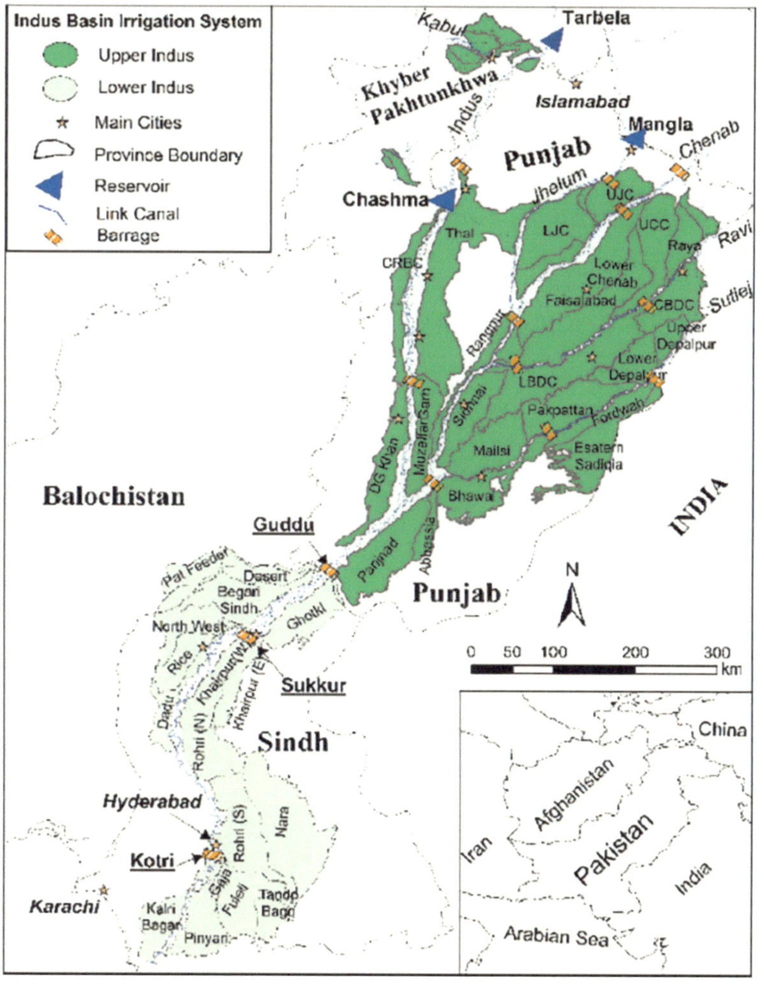

Figure 9.
IBIS with main cities, reservoirs, link canals, and barrages [29].

barrages namely: Sukkur, Guddu, and Kotri as shown in **Figure 9**. Additionally, the IBIS includes 3 major reservoirs, 12 inter-river link canals, 44 irrigation canals i.e., canal commands (14 in Sindh), 2 headworks, 2 siphons, and 16 barrages. Mostly, groundwater is utilized for the irrigation instead of canal water in regions with a lack of access to the canal water and the amount of surface water availability is insufficient [35].

3.1.2 Groundwater depth and quality

Groundwater depth and quality analyses for Sindh province were performed by the Pakistan Council of Research in Water Resources (PCRWR) [30]. The results are presented in **Figure 10**. According to PCRWR data, groundwater table depth lies between 0.2 m to 16.0 m. A major portion of Sindh's area specifically along Indus River has groundwater depth ranging from 1.6 to 3.0 m. Shallow groundwater table depth i.e., <5 m requires relatively less pumping cost and is most suitable for crop growth. It is observable that Badin, Ghotki, Khairpur, Sanghar, Shikarpur, Larkana, and Thatta lie in the range of the shallow groundwater table depth. On the other hand, most of Sindh's area has electrical conductivity (EC) greater than 4.0 dS/m. However, water having less EC is most suitable for crop growth. It is observable that the EC of water is relatively less at Badin, Ghotki, Khairpur, Sanghar, Shikarpur, Larkana, and Thatta as compared to other cities.

3.1.3 Land use land cover classification

Land use land cover (LULC) classification is a significant parameter that offers comprehensive information regarding the utilization of landscape. The LULC

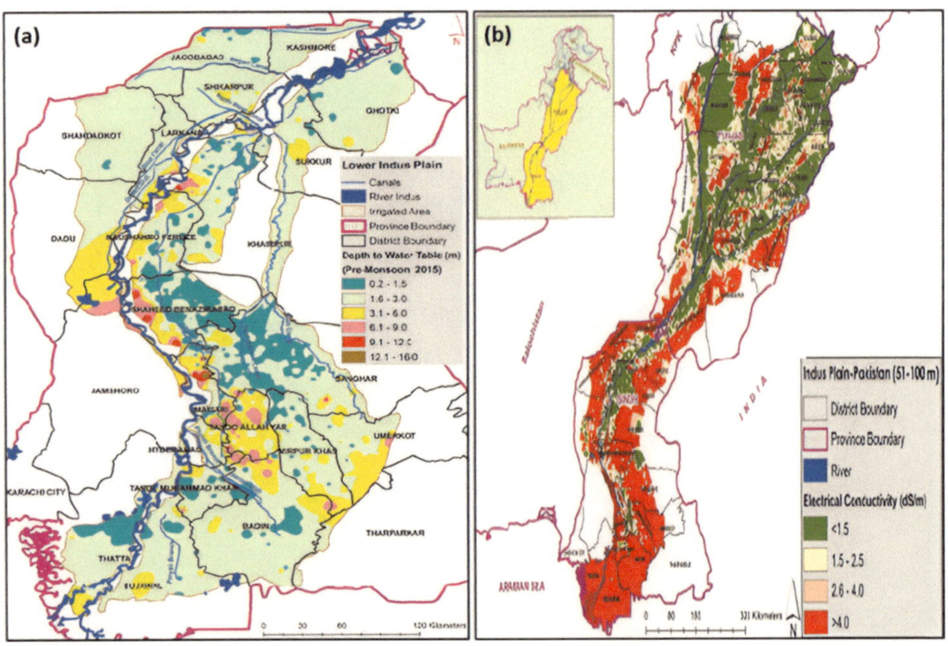

Figure 10.
Spatial variations of (a) groundwater table depth, and (b) groundwater quality for Sindh, reproduced here from [30].

classification helps in environmental monitoring allowing to identify the areas vulnerable to degradation and support in resource managements. In general, the LULC classification provides a baseline for estimating past and future changes [37]. The LULC classification was performed by Bashir et al., [31] using unsupervised classification in ArcGIS software that plays a significant role in creating LULC classifications, providing valuable information for agriculture, urban planning, environmental management, natural resource assessment, and site suitability analysis [31]. The LULC classification map is presented in **Figure 11**. It can be observed that Badin, Ghotki, Khairpur, Sanghar, Shikarpur, Larkana, and Thatta mostly lie in cropped areas/agricultural land.

3.1.4 Soil classification

Soil classification provides critical insights for appropriate cultivation practices and crops selection to improve overall yield. Ulain et al., [32] performed the soil classification for Pakistan and results are presented in **Figure 12** [32]. It is observable that, the cultivated/agricultural land of Sindh mostly experiences sandy loam, clay loam, and sandy clay loam soil. Both sandy and clay loam soils are preferable for the cultivation of food and cash crops. As per the results of the soil classification, Badin, Khairpur, Sanghar, Shikarpur, Larkana, Thatta Noshahro Feroz, Dadu, and Larkana are the most promising areas for cultivation.

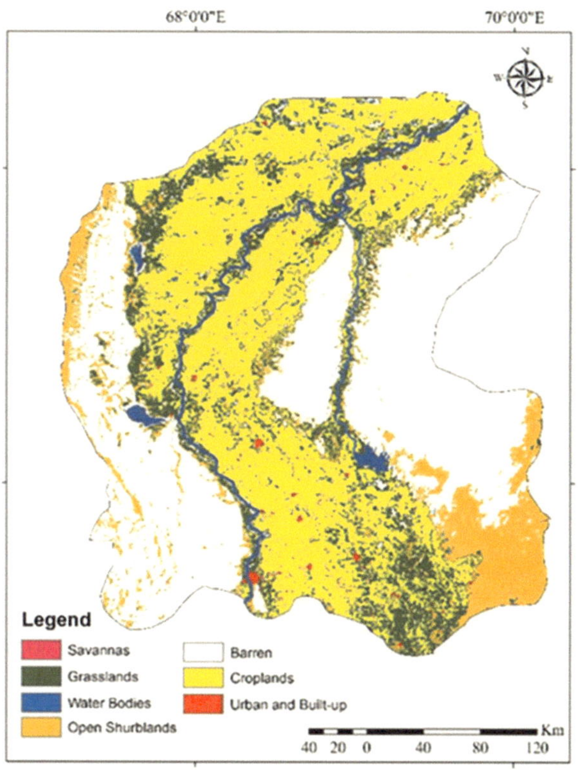

Figure 11.
Land use land cover classification of Sindh, reproduced from [31].

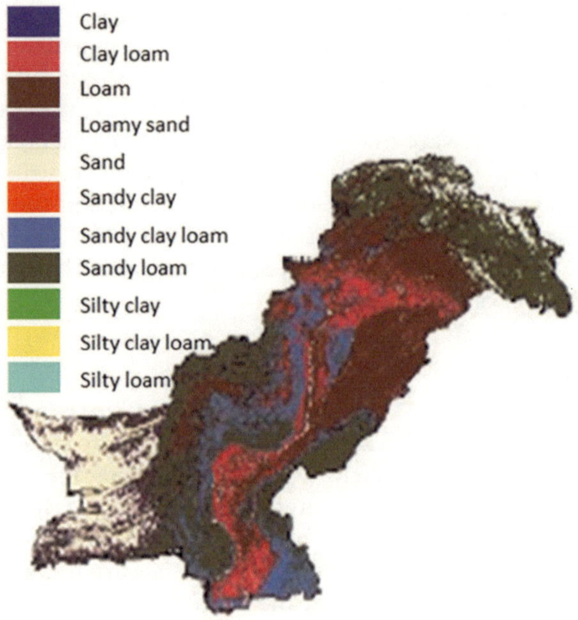

Figure 12.
Soil texture classification map for Pakistan, reproduced here from [32].

3.2 Crops selection

3.2.1 Food crops

Wheat, rice, maize, moong, jowar, gram, onion, and tomato are commonly cultivated food crops in Pakistan. **Figure 13** shows temporal variations of cultivated area for studied food crops in Pakistan and Sindh for the period of 1991–2022. The colour gradient shows temporal variations in the cultivated area for each studied food crop. In the last 32 years, the cultivated area for wheat and rice significantly changed with respect to time for Pakistan and Sindh as compared to other studied food crops. The maximum cultivated area of about 22,182 thousand acres was recorded by wheat for Pakistan and 2,920 thousand acres for Sindh during the period of 2021–2022. It is observable that wheat and rice are extensively growing crops in the country. Other food crops in Pakistan and Sindh show relatively less cultivated area trends e.g., moong, jowar, onion, and tomato.

Figure 14 shows temporal variations of net production for studied food crops in Pakistan and Sindh for the period of 1991–2022. It can be noticed that the net production trends for wheat, and rice varied significantly in Pakistan and Sindh with respect to time. During 2021–22, maximum production by wheat was recorded of 26,393.65 thousand tonnes and 3,759.75 thousand tonnes for Pakistan and Sindh, respectively.

The maximum net production of 9,322.67 thousand tonnes was recorded by rice for Pakistan and 2,861.38 thousand tonnes for Sindh. As per Pakistan's economic survey, both wheat and rice crops contributed 1.8% and 0.5% to overall GDP during 2021–22 [1].

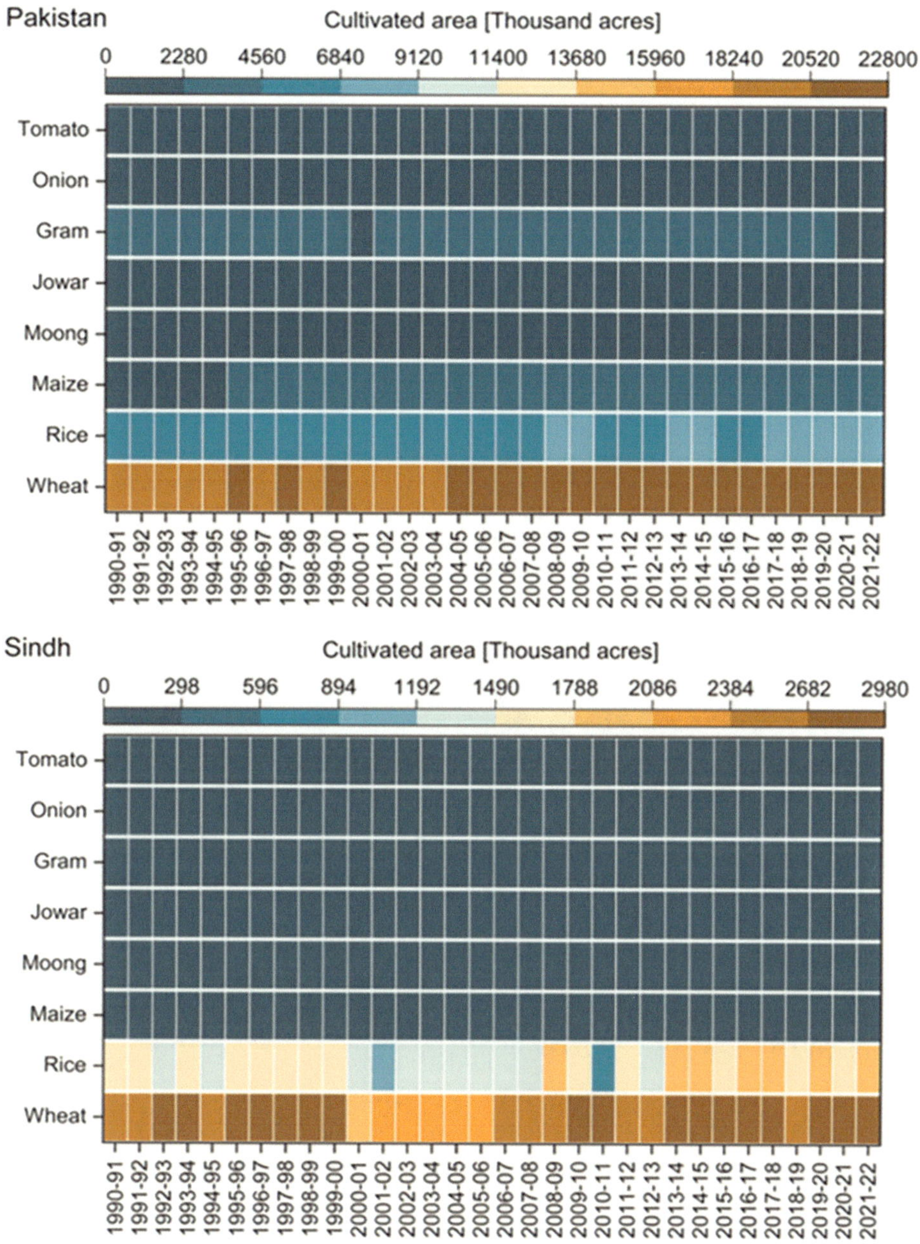

Figure 13.
Temporal variations of cultivated area for studied food crops in Pakistan and Sindh for the period of 1991–2022, data is obtained from [38].

Figure 15 shows temporal variations of the average yield for studied food crops in Pakistan and Sindh for the period of 1991–2022. In the case of food crops (wheat and rice), the maximum average yield was recorded of 1,190 kg/acre and 1,287.6 kg/acre for Pakistan and Sindh, respectively for the period of 2021–22. Similarly, the

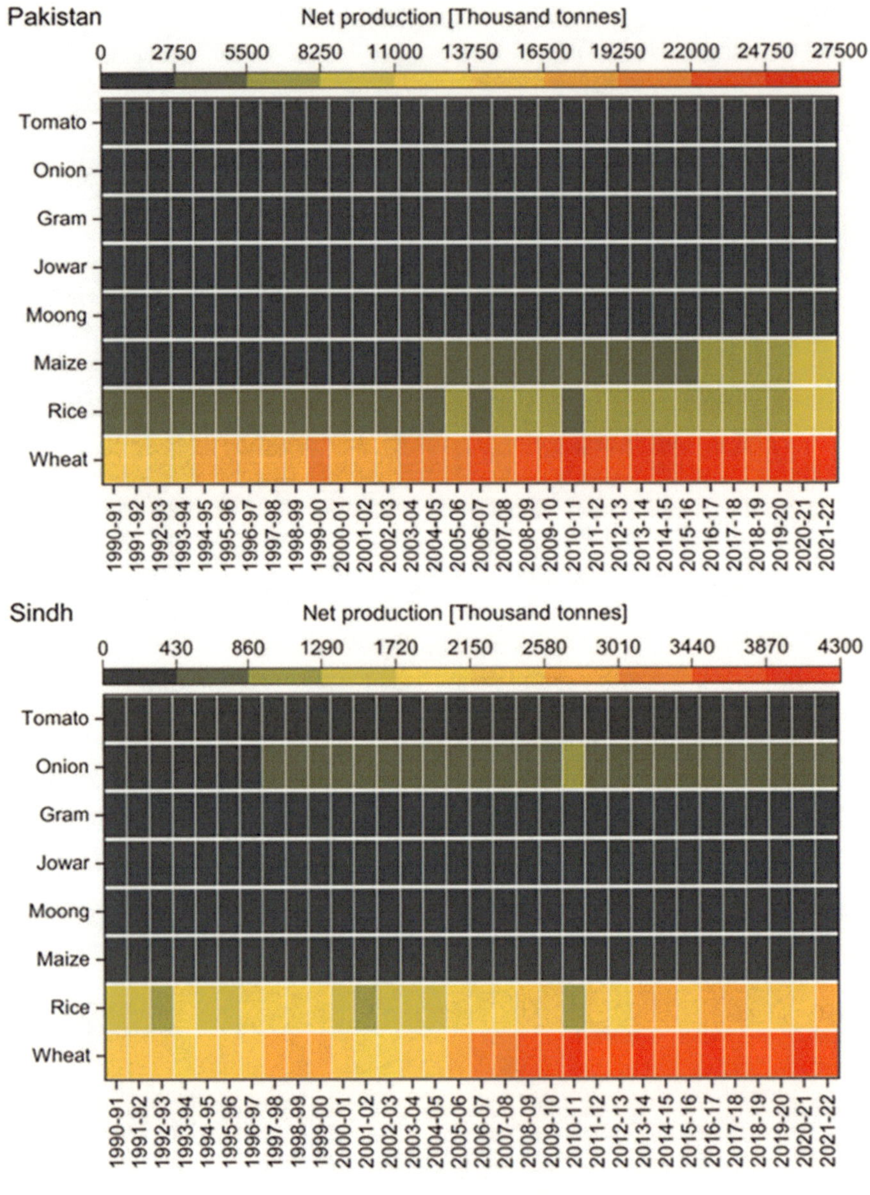

Figure 14.
Temporal variations of net production for studied food crops in Pakistan and Sindh for the period of 1991–2022, data is obtained from [38].

maximum average yield of 1,066.4 kg/acre and 1,530.4 kg/acre was recorded by rice crop for Pakistan and Sindh, respectively. On the other hand, among studied food crops maximum average yield was observed at 5,954 kg/acre and 5,285.2 by onion for both Pakistan and Sindh, respectively. Despite the high cultivated area of wheat and rice, the average yield is less as compared to developed countries because of inappropriate irrigation practices, utilization of farm machinery, and other associated operations.

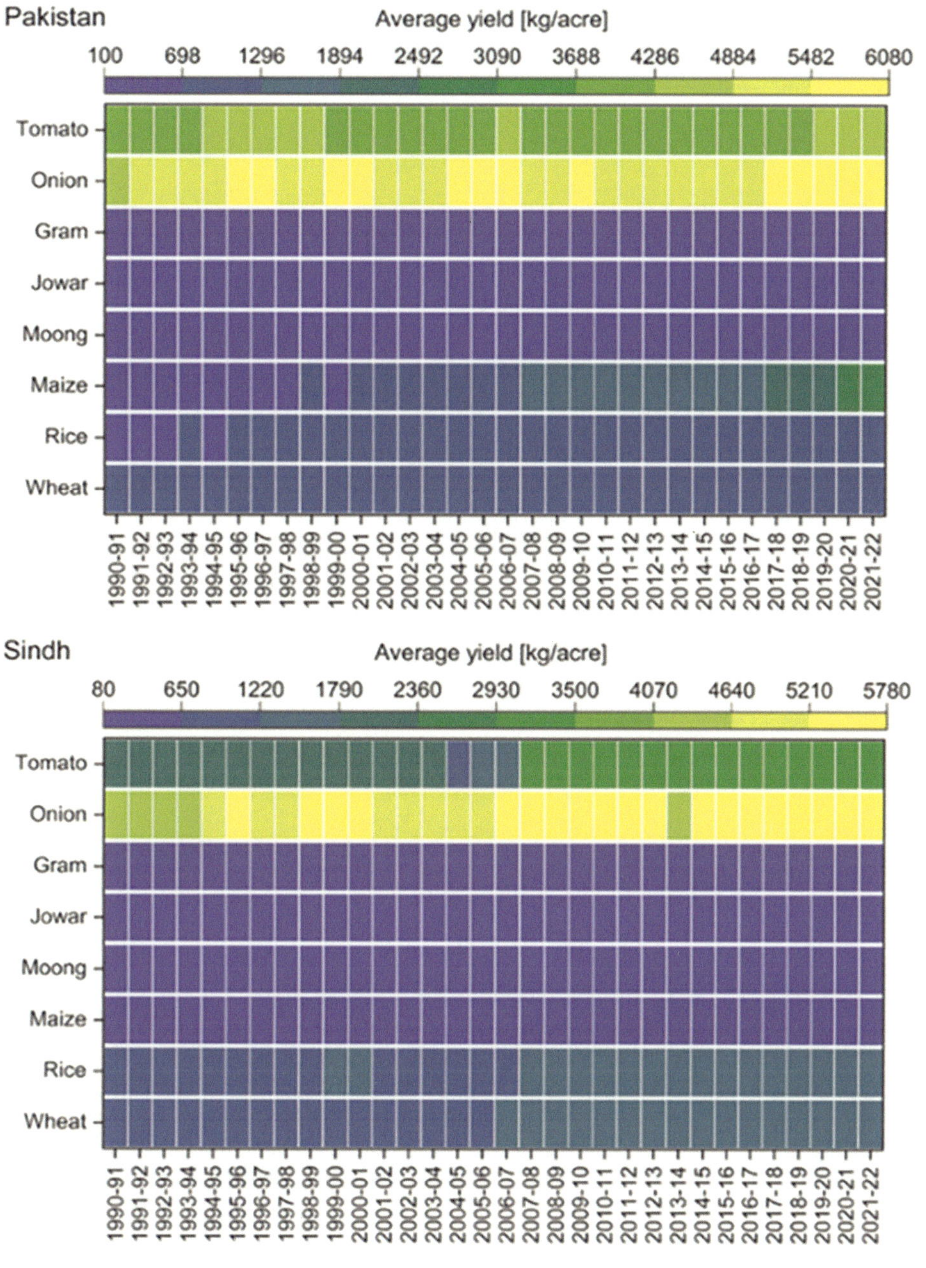

Figure 15.
Temporal variations of average yield for studied food crops in Pakistan and Sindh for the period of 1991–2022, data is obtained from [38].

Data of significant parameters for each studied food crops for the period of 2021–2022 is utilized for assigning normalized scores. The normalized scores are utilized for developing screening matrix at both country and provincial levels. **Table 1** displayed cultivated area, net production, and average yield for each studied food crop used for developing screening matrix in case of Pakistan and Sindh for the period of

Food crop	Cultivated area (Thousand acres)	Net production (Thousand tonnes)	Average yield (kg/acre)
		Pakistan	
Wheat	22,181.77	26,393.65	1,190
Rice	8,741.19	9,322.67	1,066.4
Maize	4,083.50	10,634.83	2,604.4
Moong	745.66	263.78	353.6
Jowar	189.11	63.94	338
Gram	2,142.46	319.32	149.2
Onion	348.42	2,074.47	5,954
Tomato	124.22	541.28	4,357.2
		Sindh	
Wheat	2,920.13	3,759.75	1,287.6
Rice	1,869.6	2,861.38	1,530.4
Maize	10.51	4.37	416
Moong	32.35	5.49	169.6
Jowar	25.81	9.91	384
Gram	59.00	22.88	388
Onion	156.41	826.69	5,285.2
Tomato	59.58	176.90	2,969.2

Table 1.
Cultivated area, net production, and average yield for each studied food crop used for developing screening matrix in case of Pakistan and Sindh for the period of 2021–2022 [38].

2021–2022. **Table 2** shows case data of irrigation requirements, water cost, production cost, and net profit of each studied food crop used for developing a screening matrix in case of Pakistan and Sindh. It is worth mentioning that the data of irrigation requirements, water cost, production cost, and net profit of each studied food crops

Food crop	Irrigation requirements (mm/season)	Water cost (Rs/acre)	Production cost (Rs/acre)	Net profit (Rs/acre)
Wheat	325–450	3,109	27,432	38,970
Rice	450–700	14,995	46,773	40,122
Maize	500–800	6,537	49,672	38,187
Moong	300–400	1,963	19,158	26,509
Jowar	450–600	2,500	30,035	15,062
Gram	300–450	299	8,163	6,575
Onion	350–550	7,351	38,734	37,950
Tomato	400–800	4,915	74,699	61,601

Table 2.
Irrigation requirements, water cost, production cost, and net profit of each studied food crop used for developing screening matrix in case of Pakistan and Sindh for the period of 2021–2022 [38].

are used for Punjab province as per availability. Therefore, the data is used as a reference for developing a screening matrix for both Pakistan and Sindh province.

Normalized scores are computed based on each parameter and stars are assigned corresponding to these scores. **Figure 16** shows normalized score cards with cumulative stars obtained by each studied food crop for Pakistan and Sindh. Maximum stars obtained by the studied food and cash crops are selected as potential food crops. Among studied food crops, wheat and tomato crops secured maximum stars in case of Pakistan whereas wheat and rice crops secured maximum stars in case of Sindh. Therefore, wheat and rice crops are selected as potential food crops for Sindh province.

Pakistan	Wheat	Rice	Maize	Moong	Jowar	Gram	Onion	Tomato
Cultivated area [Thousand acre]	1.00	0.39	0.18	0.03	0.01	0.09	0.01	0.00
Irrigation req. [mm/season]	0.13	0.67	1.00	0.00	0.58	0.08	0.33	0.83
Water cost [Rs/acre]	0.81	0.00	0.58	0.89	0.85	1.00	0.52	0.69
Net production [Thousand tonnes]	1.00	0.35	0.40	0.01	0.00	0.01	0.08	0.02
Average yield [kg/acre]	0.18	0.16	0.42	0.04	0.03	0.00	1.00	0.72
Production cost [Rs./acre]	0.71	0.42	0.38	0.83	0.67	1.00	0.54	0.00
Net profit [Rs./acre]	0.59	0.61	0.57	0.36	0.15	0.00	0.57	1.00
Cumulative stars	24	17	19	16	16	15	18	21

Sindh	Wheat	Rice	Maize	Moong	Jowar	Gram	Onion	Tomato
Cultivated area [Thousand acre]	1.00	0.64	0.00	0.01	0.01	0.02	0.05	0.02
Irrigation req. [mm/season]	0.13	0.67	1.00	0.00	0.58	0.08	0.33	0.83
Water cost [Rs/acre]	0.81	0.00	0.58	0.89	0.85	1.00	0.52	0.69
Net production [Thousand tonnes]	1.00	0.76	0.00	0.00	0.00	0.01	0.22	0.05
Average yield [kg/acre]	0.22	0.27	0.05	0.04	0.04	0.04	1.00	0.55
Production cost [Rs./acre]	0.71	0.42	0.38	0.83	0.67	1.00	0.54	0.00
Net profit [Rs./acre]	0.59	0.61	0.57	0.36	0.15	0.00	0.57	1.00
Cumulative stars	25	22	16	16	16	15	19	20

Figure 16.
Normalized score cards from viewpoints of significant factors of each studied food crop with cumulative stars for Pakistan and Sindh.

3.2.2 Cash crops

Cotton, sugarcane, rapeseed, chilies, sesame, potato, mango, and dates are commonly cultivated cash crops in Pakistan. **Figure 17** shows temporal variations of cultivated area for studied cash crops in Pakistan and Sindh for the period of 1991–2022. In the last 32 years, the cultivated area for cotton and sugarcane crops significantly changed with time for Pakistan and Sindh as compared to other studied cash crops.

Among studied cash crops, a maximum cultivated area of about 4,786.44 thousand acres and 1,467.56 thousand acres was recorded by cotton crops for Pakistan and

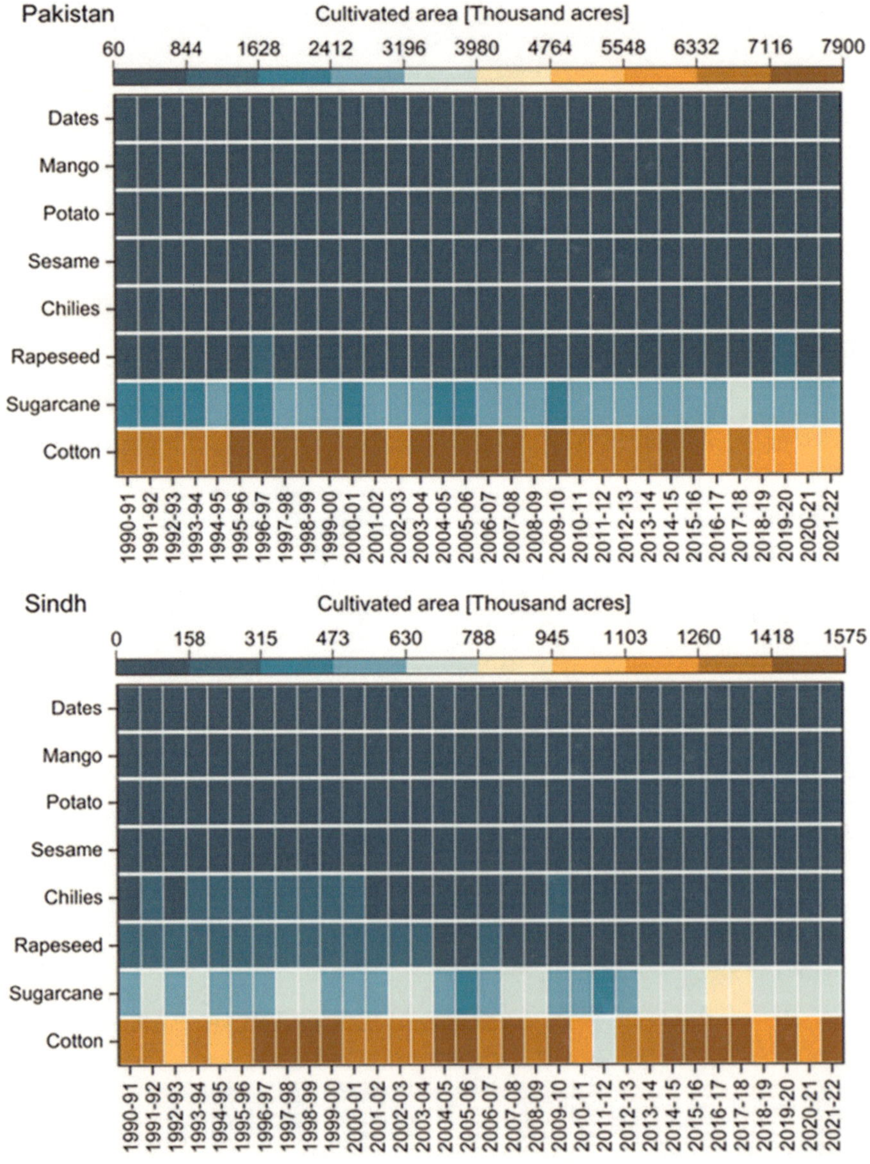

Figure 17.
Temporal variations of cultivated area for studied cash crops in Pakistan and Sindh for the period of 1991–2022, data is obtained from [38].

Sindh, respectively. Similarly, a maximum cultivated area of 3,114.31 thousand acres and 729.58 thousand acres was recorded by sugarcane crops in both Pakistan and Sindh. It is observable that cotton and sugarcane crops are extensively growing cash crops in Pakistan.

Figure 18 shows temporal variations of net production for studied cash crops in Pakistan and Sindh for the period of 1991–2022. It can be noticed that the net production trends for cotton, and sugarcane crops varied significantly in Pakistan and Sindh with respect to time.

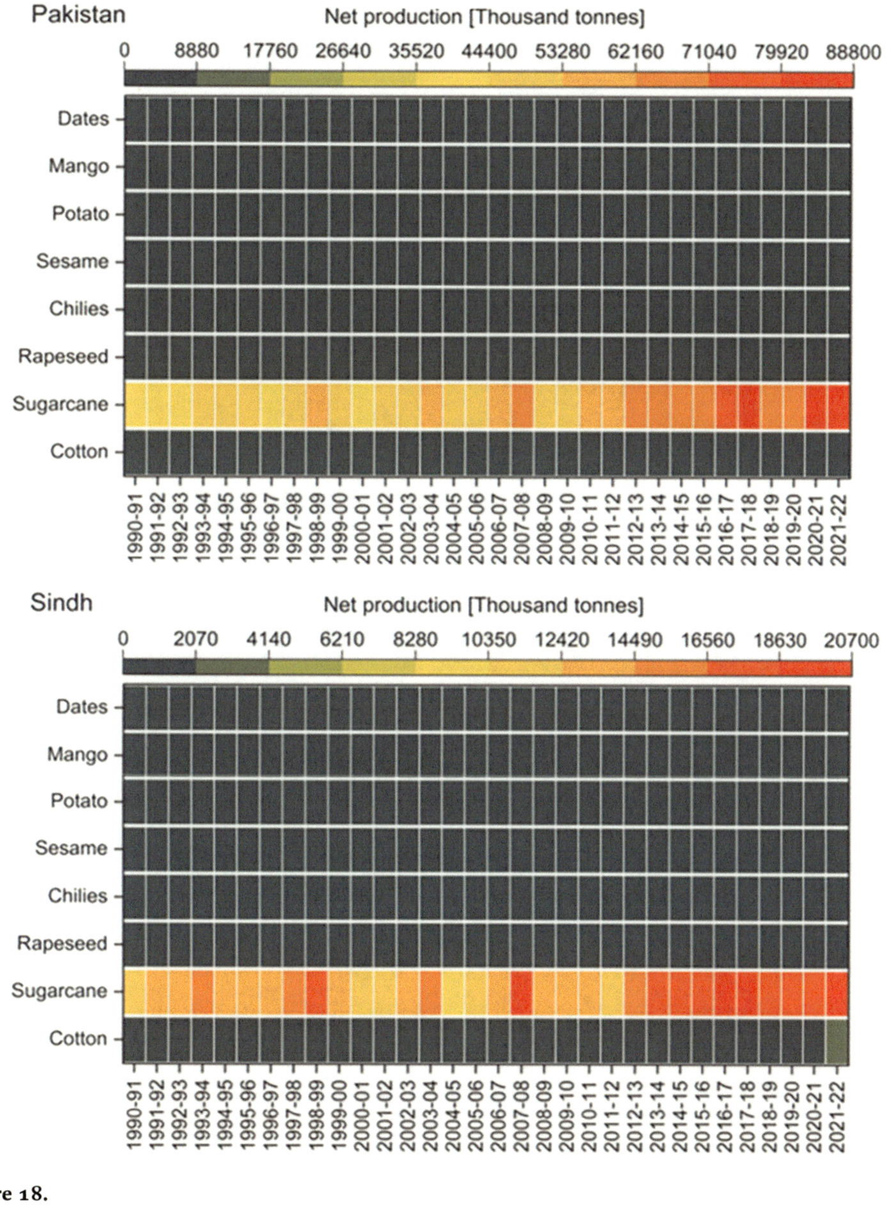

Figure 18.
Temporal variations of net production for studied cash crops in Pakistan and Sindh for the period of 1991–2022, data is obtained from [38].

During 2021–22, maximum net production of 8,328.81 thousand tonnes and 2,998.41 thousand tonnes by cotton for Pakistan and Sindh, respectively. Likewise, the maximum net production of 88,650.59 thousand tonnes, and 19,460.72 thousand tonnes was observed by sugarcane crops for Pakistan and Sindh, respectively. During 2021–2022, both cotton and sugarcane crops contributed 0.6% and 0.8% to overall GDP of the country as reported by Pakistan Economic Survey [1].

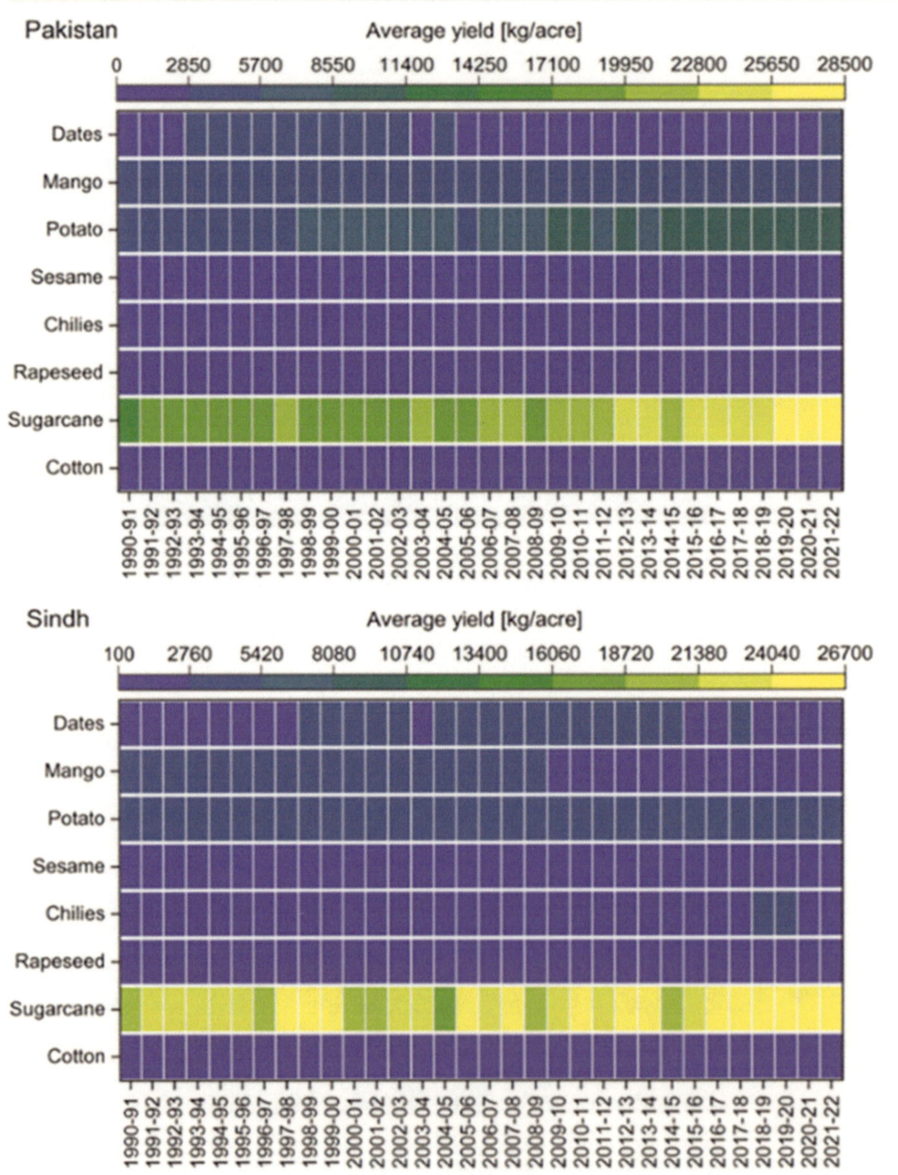

Figure 19.
Temporal variations of average yield for studied cash crops in Pakistan and Sindh for the period of 1991–2022, data is obtained from [38].

DOI: http://dx.doi.org/10.5772/intechopen.114144

Figures 15, 16 and **19** shows temporal variations of the average yield for studied cash crops in Pakistan and Sindh for the period of 1991–2022. In case of cash crops (cotton and sugarcane), the maximum average yield was recorded of 1,740 kg/acre and 1241.2 kg/acre by cotton crop for Pakistan and Sindh, respectively. The maximum average yield was recorded of 28,465.6 kg/acre and 26,673.6 kg/acre by sugarcane crops for both Pakistan and Sindh, respectively. Despite the high cultivated area of cotton and sugarcane in Pakistan and Sindh, the average yield is relatively less as compared to other developed countries because of inappropriate irrigation practices, utilization of farm machinery and other associated operations.

Table 3 displays cultivated area, net production, and average yield for each studied cash crop used for developing screening matrix in case of Pakistan and Sindh for the period of 2021–2022. **Table 4** shows case data of irrigation requirements, water cost, production cost, and net profit of each studied cash crop used for developing screening matrix in case of Pakistan and Sindh. **Figure 20** shows normalized score cards with cumulative stars obtained by each studied cash crop. Among studied cash crops, cotton and sugarcane crops secured maximum stars at both the country and provincial level. Therefore, cotton and sugarcane crops are selected as potential cash crops for Sindh province.

Cash crop	Cultivated area (Thousand acres)	Net production (Thousand tonnes)	Average yield (kg/acre)
		Pakistan	
Cotton	4,786.44	8,328.81	1,740
Sugarcane	3,114.31	88,650.59	28,465.6
Rapeseed	812.91	454.52	559.2
Chilies	143.6	144.0	1,004
Sesame	493.95	127.98	259.2
Potato	775.32	7,931.07	10,229.6
Mango	393.52	1,844.71	4,687.6
Dates	238.50	838.25	3,514.8
		Sindh	
Cotton	1,467.56	2,998.41	2,043.2
Sugarcane	729.58	19,460.72	26,673.6
Rapeseed	134.27	52.66	392.4
Chilies	109.6	111.4	1016
Sesame	34.42	6.31	183.6
Potato	1.50	6.00	3,987.6
Mango	146.19	387.40	2,650
Dates	101.38	224.03	2,210

Table 3.
Cultivated area, net production, and average yield for each studied cash crop used for developing screening matrix in case of Pakistan and Sindh for the period of 2021–2022 [38].

Cash crop	Irrigation requirements (mm/season)	Water cost (Rs/acre)	Production cost (Rs/acre)	Net profit (Rs/acre)
Cotton	1,055–1,655	6,349	46,830	32,647
Sugarcane	2,375–3,108	17,751	77,044	70,929
Rapeseed	400–600	1,571	22,535	42,596
Chilies	400–600	8,888	73,061	63,888
Sesame	350–450	1,992	40,335	2,490
Potato	500–700	6,819	99,846	168,492
Mango	500–800	9,000	203,762	87,488
Dates	1,055–1,455	5,000	46,830	32,647

Table 4.
Irrigation requirements, water cost, production cost, and net profit of each studied cash crop used for developing screening matrix in case of Pakistan and Sindh for the period of 2021–2022 [38].

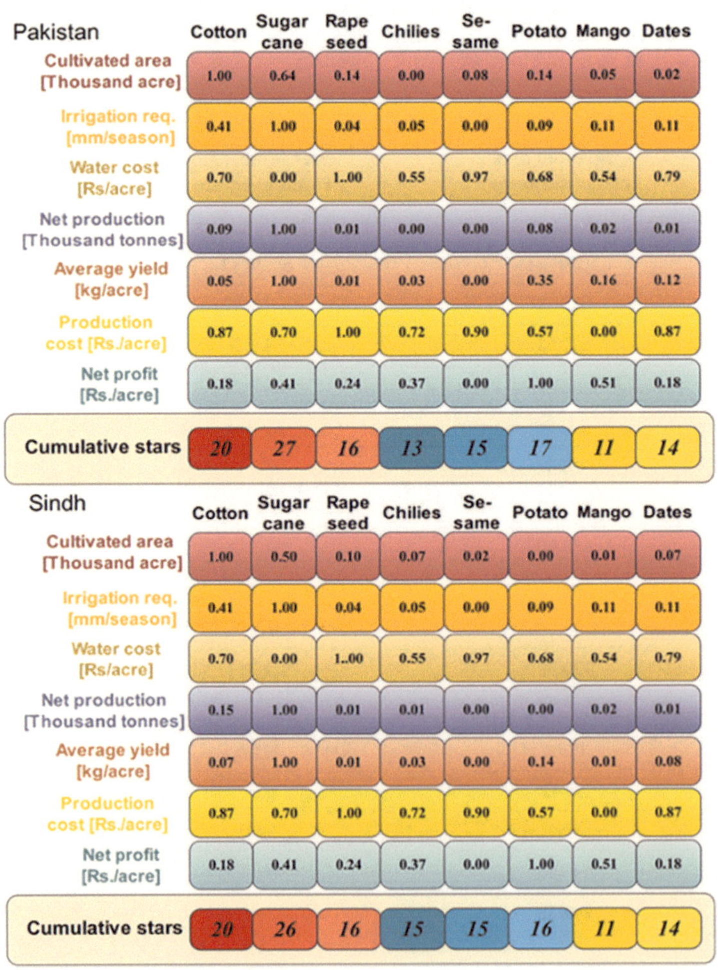

Figure 20.
Normalized score cards from viewpoints of significant factors of each studied cash crop with cumulative stars for Pakistan and Sindh.

4. Conclusion

IBIS is the primary source for meeting irrigation requirements of Pakistan specifically for Sindh. However, irrigation practices in Pakistan are constrained by poor irrigation scheduling or inappropriate estimation of actual crop water requirements corresponds to crop's growth and development stages resulting in high water consumption and lower crop yields. Furthermore, extensive pumping for irrigation consumes a significant amount of primary energy as well as fresh water. In this regard, hybrid energy powered smart irrigation system (HEPSIS) is an emerging solution for increasing crop yield by optimizing both energy and water. However, site suitability and crops selection analyses are essential before design/development, installation, and testing of the smart irrigation system. Therefore, the study aims to explore suitable installation site and crops for Sindh province. Site suitability analyses involves IBIS mapping, groundwater table depth/quality mapping, land use land cover, and soil classifications. For crops selection, eight kinds of food and cash crops are investigated by using a qualitative methodology based on stars i.e., screening matrix approach by considering potential parameters including cultivated area, irrigation requirements water cost, net production, average yield, production cost, and net profit. Normalized scoring method is utilized for assigning stars to each studied food and cash crop based on their potential parameters. According to results, Badin, Ghotki, Khairpur, Sanghar, Shikarpur, Larkana, and Thatta are selected as some suitable sites for the HEPSIS. Additionally, among studied crops, wheat and rice are selected as potential food crops while cotton and sugarcane are selected as potential cash crops which will be irrigated with smart irrigation system.

Acknowledgements

The authors would like to express their gratitude to Innovate UK's Energy Catalyst programme (Ayrton Funding provided by the Foreign, Commonwealth Development Office through their Transforming Energy Access Programme) and UK aid for the support and funding provided for the research project titled 'Hybrid Energy Powered Smart Irrigation System for Smallholder Farmers' (Project number: 10039507). This financial support from Innovate UK's Energy Catalyst programme has played a pivotal role in facilitating the successful execution of our research as well as for the completion of this book.

Conflict of interest

The authors declare no conflict of interest.

Author details

Muhammad Aleem[1†], Muhammad Sultan[1*†], Muhammad Imran[2†], Zafar A. Khan[3], Hadeed Ashraf[1], Hafiz M. Asfahan[1] and Fiaz Ahmad[1]

1 Department of Agricultural Engineering, Bahauddin Zakariya University, Multan, Pakistan

2 Department of Mechanical, Biomedical and Design Engineering, College of Engineering and Physical Sciences, Aston University, Birmingham, United Kingdom

3 Department of Electrical Engineering, Mirpur University of Science and Technology, Mirpur A.K., Pakistan

*Address all correspondence to: muhammadsultan@bzu.edu.pk

† These authors contributed equally.

References

[1] Finance Division Government of Pakistan. Pakistan Economic Survey 2021–2022. Islamabad, Pakistan: Ministry of Finance; 2022

[2] Aleem M, Sultan M, Farooq M, Riaz F, Yakout SM, Ahamed MS, et al. Evaluating the emerging adsorbents for water production potential and thermodynamic limits of adsorption-based atmospheric water harvesting systems. International Communications in Heat and Mass Transfer. 2023;**145**:106863. DOI: 10.1016/j.icheatmasstransfer.2023.106863

[3] Wasti TZ, Sultan M, Aleem M, Sajjad U, Farooq M, Raza HMU, et al. An overview of solid and liquid materials for adsorption-based atmospheric water harvesting. Advanced Mechanical Engineering. 2022;**14**: 16878132221082768. DOI: 10.1177/16878132221082768

[4] Aleem M, Sultan M, Mahmood MH, Miyazaki T. In: Sultan M, Miyazaki T, editors. Desiccant Dehumidification Cooling System for Poultry Houses in Multan (Pakistan) BT - Energy-Efficient Systems for Agricultural Applications. Cham: Springer International Publishing; 2022. pp. 19-42

[5] Syed A, Raza T, Bhatti TT, Eash NS. Climate impacts on the agricultural sector of pakistan: Risks and solutions. Environmental Challenges. 2022;**6**: 100433. DOI: 10.1016/j.envc.2021.100433

[6] Mujtaba A, Nabi G, Masood M, Iqbal M, Asfahan HM, Sultan M, et al. Impact of cropping pattern and climatic parameters in lower Chenab Canal System—Case study from Punjab Pakistan. Agriculture. 2022;**12**:708

[7] Ismail M, Ahmed E, Peng G, Xu R, Sultan M, Khan FU, et al. Evaluating the impact of climate change on the stream flow in Soan River Basin (Pakistan). Water. 2022;**14**. DOI: 10.3390/w14223695

[8] Ahamed MS, Sultan M, Shamshiri RR, Rahman MM, Aleem M, Balasundram SK. Present status and challenges of fodder production in controlled environments: A review. Smart Agricultural Technology. 2023;**3**: 100080. DOI: 10.1016/j.atech.2022.100080

[9] Ali A, Hussain T, Tantashutikun N, Hussain N, Cocetta G. Application of smart techniques, internet of things and data mining for resource use efficient and sustainable crop production. Agriculture. 2023;**13**. DOI: 10.3390/agriculture13020397

[10] Hussain G, Aleem M, Sultan M, Sajjad U, Ibrahim SM, Shamshiri RR, et al. Evaluating evaporative cooling assisted solid desiccant dehumidification system for agricultural storage application. Sustainability. 2022;**14**. DOI: 10.3390/su14031479

[11] Aziz M, Khan M, Anjum N, Sultan M, Shamshiri RR, Ibrahim SM, et al. Scientific irrigation scheduling for sustainable production in olive groves. Agriculture. 2022;**12**. DOI: 10.3390/agriculture12040564

[12] Solangi GS, Shah SA, Alharbi RS, Panhwar S, Keerio HA, Kim T-W, et al. Investigation of irrigation water requirements for major crops using CROPWAT model based on climate data. Water. 2022;**14**:2578. DOI: 10.3390/w14162578

[13] Asfahan HM, Sultan M, Ahmad F, Majeed F, Ahamed MS, Aziz M, et al. . In: Sultan M, Ahmad F, editors. Agrovoltaic and Smart Irrigation:

Pakistan Perspective. London, UK, Rijeka: IntechOpen; 2022

[14] Siyal AW, Gerbens-Leenes PW. The water–energy nexus in irrigated agriculture in South Asia: Critical hotspots of irrigation water use, related energy application, and greenhouse gas emissions for wheat, rice, sugarcane, and cotton in Pakistan. Water. 2022;**4**. DOI: 10.3389/frwa.2022.941722

[15] Tagar A, Chandio FA, Mari IA, Wagan B. Comparative study of drip and furrow irrigation methods at farmer-s field in Umarkot. International Journal of Agricultural and Biosystem Engineering. 2012;**6**:788-792

[16] Mahmood A, Oweis T, Ashraf M, Majid A, Aftab M, Aadal NK, et al. Performance of improved practices in farmers' fields under rainfed and supplemental irrigation systems in a semi-arid area of Pakistan. Agricultural Water Management. 2015;**155**:1-10. DOI: 10.1016/j.agwat.2015.03.006

[17] Aziz M, Rizvi SA, Iqbal MA, Syed S, Ashraf M, Anwer S, et al. A sustainable irrigation system for small landholdings of Rainfed Punjab, Pakistan. Sustainability. 2021;**13**:11178

[18] Razzaq A, Rehman A, Qureshi AH, Javed I, Saqib R, Iqbal MN. An economic analysis of high efficiency irrigation systems in Punjab, Pakistan. Sarhad Journal of Agriculture. 2018;**34**:818-826

[19] Cáceres G, Millán P, Pereira M, Lozano D. Smart farm irrigation: Model predictive control for economic optimal irrigation in agriculture. Agronomy. 2021;**11**. DOI: 10.3390/agronomy11091810

[20] Pereira GP, Chaari MZ, Daroge F. IoT-enabled smart drip irrigation system using ESP32. IoT. 2023;**4**:221-243

[21] Obaideen K, Yousef BAA, AlMallahi MN, Tan YC, Mahmoud M, Jaber H, et al. An overview of smart irrigation systems using IoT. Energy Nexus. 2022;**7**:100124. DOI: 10.1016/j.nexus.2022.100124

[22] Hassan ES. Energy-efficient resource allocation algorithm for CR-WSN-based smart irrigation system under realistic scenarios. Agriculture. 2023;**13**. DOI: 10.3390/agriculture13061149

[23] Sami M, Khan SQ, Khurram M, Farooq MU, Anjum R, Aziz S, et al. A deep learning-based sensor modeling for smart irrigation system. Agronomy. 2022;**12**. DOI: 10.3390/agronomy12010212

[24] Ahmad U, Alvino A, Marino S. Solar fertigation: A sustainable and smart IoT-based irrigation and fertilization system for efficient water and nutrient management. Agronomy. 2022;**12**. DOI: 10.3390/agronomy12051012

[25] Campos GS, Rocha AR, Gondim R, Coelho da Silva TL, Gomes DG. Green: An Internet-of-Things framework for smart irrigation. Sensors. 2020;**20**. DOI: 10.3390/s20010190

[26] Vallejo-Gómez D, Osorio M, Hincapié CA. Smart irrigation systems in agriculture: A systematic review. Agronomy. 2023;**13**. DOI: 10.3390/agronomy13020342

[27] Bwambale E, Abagale FK, Anornu GK. Smart irrigation monitoring and control strategies for improving water use efficiency in precision agriculture: A review. Agricultural Water Management. 2022;**260**:107324. DOI: 10.1016/j.agwat.2021.107324

[28] Mallareddy M, Thirumalaikumar R, Balasubramanian P, Naseeruddin R, Nithya N, Mariadoss A, et al.

Maximizing water use efficiency in rice farming: A comprehensive review of innovative irrigation management technologies. Water. 2023;**15**. DOI: 10.3390/w15101802

[29] Van Steenbergen F, Basharat M, Lashari BK. Key challenges and opportunities for conjunctive management of surface and groundwater in mega-irrigation systems: Lower Indus, Pakistan. Resources. 2015; **4**:831-856

[30] Iqbal N, Ashraf M, Imran M, Salam HA, Hasan FU, Khan AD. Groundwater Investigations and Mapping in the Lower Indus Plain. Islamabad: Pakistan Council of Research in Water Resources; 2020

[31] Bashir B, Cao C, Naeem S, Zamani Joharestani M, Bo X, Afzal H, et al. Spatio-temporal vegetation dynamic and persistence under climatic and anthropogenic factors. Remote Sensors. 2020;**12**. DOI: 10.3390/rs12162612

[32] Ulain Q, Ali SM, Shah AA, Iqbal KM, Ullah W, Tariq MA. Identification of soil erosion-based degraded land areas by employing a geographic information system: A case study of Pakistan for 1990–2020. Sustainability. 2022;**14**. DOI: 10.3390/su141911888

[33] Ahmad MI, Oxley L, Ma H. What makes farmers exit farming: A case study of Sindh Province, Pakistan. Sustainability. 2020;**12**. DOI: 10.3390/su12083160

[34] Directorate of Agriculture (Economics & Marketing), L. Agriculture Marketing Information Service (AMIS). Available from: http://amis.pk/

[35] Muzammil M, Zahid A, Breuer L. Water resources management strategies for irrigated agriculture in the Indus Basin of Pakistan. Water. 2020;**12**. DOI: 10.3390/w12051429

[36] Lytton L, Ali A, Garthwaite B, Punthakey JF, Saeed B. Groundwater in Pakistan's Indus Basin: Present and future Prospects. Washington, DC: World Bank; 2021

[37] Arulbalaji P, Padmalal D, Sreelash K. GIS and AHP techniques based delineation of groundwater potential zones: A case study from Southern Western Ghats, India. Scientific Reports. 2019;**9**:2082. DOI: 10.1038/s41598-019-38567-x

[38] Directorate of Agriculture (Economics & Marketing), L. (Punjab) Agriculture Marketing Information Service (AMIS). Available from: http://www.amis.pk/

Chapter 2

The Evolution of Sustainable Rice Production along the Lower Mississippi River (USA) with the Increasing Spector of Climate Change

Michael Aide

Abstract

It is incumbent to assess the status of U.S. rice production and its agronomic practices and then predict if the status is sustainable with climate change. Climate change expectations include a strong likelihood of higher temperatures and some uncertainty in precipitation. Technological solutions appear to be revolving around (i) rice breeding to improve cultivar heat tolerance, especially for high night-time temperatures, and (ii) altering irrigation regimes to conserve groundwater. Of concern are the potential of protracted droughts in the Sacramento Valley of California, salinization along the gulf coast and aquifer depletion across portions of the mid-South. The objectives of this manuscript include: (i) evaluate existing US Mid-South rice irrigation strategies, (ii) assess the yield potential and seed quality of emerging water-conserving irrigation regimes, (iii) assess the influence of rice irrigation regimes on aquifer overdraft, and (iv) assess the influence of increased temperatures on rice growth and development. Alternate wetting and drying irrigation and furrow irrigation are attractive irrigation regimes to reduce aquifer depletion. Yield trials show mixed results, with yield differences associated with soil type, timing and frequency of irrigations, nitrogen fertilization, and variety selection. Producer acceptance of furrow irrigation is rapidly developing, even in rice producing regions that have not experienced aquifer overdraft.

Keywords: *Oryza sativa*, aquifers, water conservation, wetting and drying irrigation, furrow irrigation

1. Introduction

Four regions in the United States support rice (*Oryza sativa* L) production: (i) Arkansas Grand Prairie, (ii) the Mississippi Delta along the lower Mississippi River, (iii) the eastern Texas and southwest Louisiana Gulf Coast, and (iv) the Sacramento

Valley in California. Each of these regions specializes in rice production using unique production methods and specific rice types. Rice types are referred to by the ratio of grain length to grain width as long grain, medium grain, and short grain rice. California produces primarily medium grain rice and constitutes approximately 20% of the US rice harvest. The lower Mississippi River valley and the Gulf Coast producing areas primarily produce long-grain rice, with some new interest in medium grain rice. Short grain rice is a minor entity.

The Mid-South rice producing region is located along the Mississippi River in Missouri, Arkansas, Mississippi, and Louisiana. The state of Arkansas typically grows most of the U.S. long-grain crop; however, increasing quantities of medium-grain rice are currently cultured. All U.S. rice is produced in irrigated fields that are experiencing increasing instances of spikelet sterility because of high nighttime temperatures, delayed planting because of changing rainfall patterns, aquifer depletion, and surface water scarcity, all of which results in altered plant physiology, nutrient uptake, and net photosynthesis.

In this manuscript, the mid-South region along the Mississippi River will be emphasized. The objectives of this manuscript include: (i) evaluate existing US Mid-South rice irrigation strategies (delayed flood), (ii) assess the yield potential and seed quality of emerging water-conserving irrigation regimes compared to traditional rice production systems, (iii) assess the ability of alternate wetting and drying irrigation and furrow irrigation on limiting aquifer overdraft, and (iv) assess the influence of increased temperatures on rice growth and development.

2. Rice production in the Mid-South USA

Most of the rice production in Mississippi, Arkansas and Missouri is in the lower Mississippi River embayment and presents itself as a fluvial landscape. Texas and Louisiana's rice production is mainly located on the coastal plain and smaller coastal floodplains. The east-central Arkansas Grand Prairie Region is also a major rice producing region. From a global perspective, rice production in the U.S. is not a dominant crop; however, rice production does contribute to the global export capacity. U.S. rice production centers around 1.1 million ha with an average recent annual yield near 8300 kg ha^{-1} (**Table 1**). Prior to milling, annual Mid-South rice production

State	Production (million cwt)	Yield (kg ha^{-1})	Value (million $)
Arkansas	80.3	8305	1373
Louisiana	27.5	7398	473
California	22.3	9819	752
Texas	12.1	7297	207
Missouri	11.8	8900	192
Mississippi	6.2	8261	105

Source: Data from https://www.nass.usda.gov (USDA National Agricultural Statistics Service) [1]; CWT is centum weight or 100 lb. or 50.8 kg.

Table 1.
Leading rice producing U.S. States in 2022.

Type	2022	2021	2000	2019	2018
Long grain	128.2	144.5	170.8	125.6	163.6
Short grain	30.0	44.3	53.9	57.1	57.3

Source: Data from https://www.nass.usda.gov (USDA National Agricultural Statistics Service) [1].

Table 2.
U.S. rice production over time for long/medium grain (million cwt).

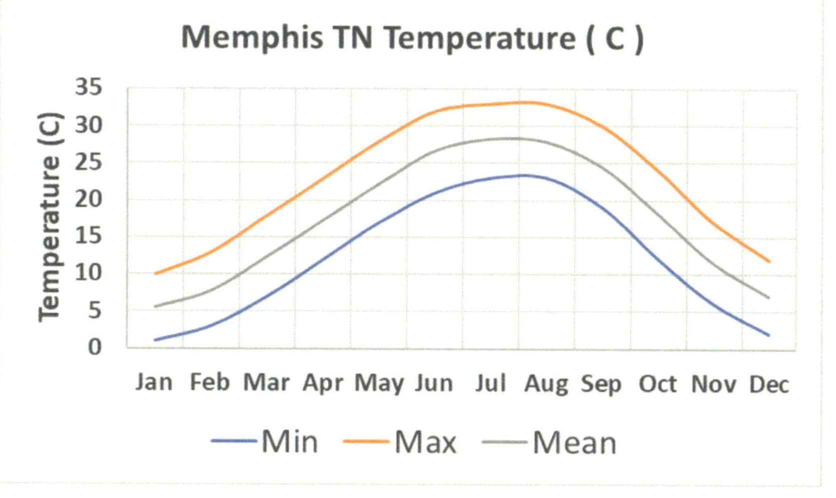

Figure 1.
The average, minimum and maximum temperature by month at Memphis TN. Source: Data from National Weather Service [2] at https://www.weather.gov/media/meg/August2017ClimateSummary.pdf.

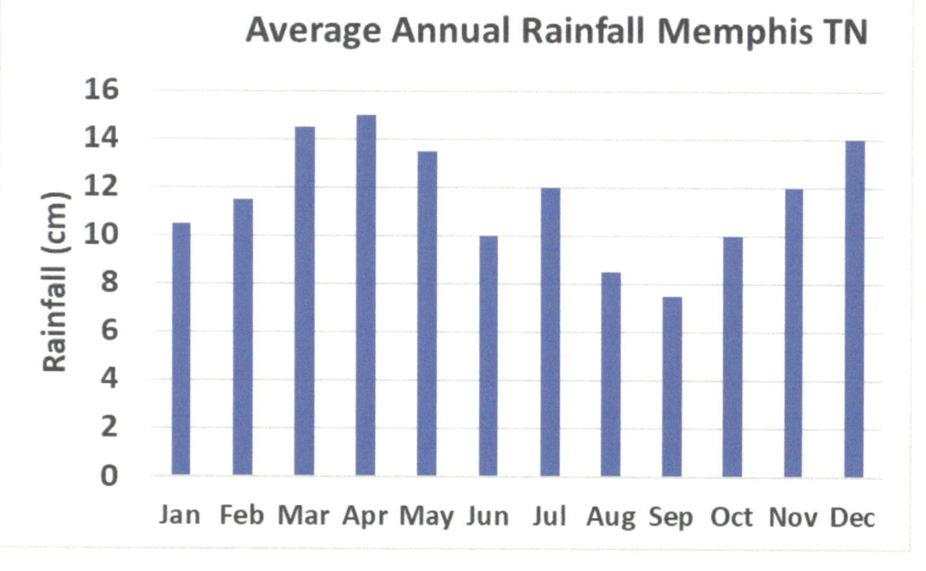

Figure 2.
The mean rainfall amounts by month at Memphis TN. Source: Data from National Weather Service [2] at https://www.weather.gov/media/meg/August2017ClimateSummary.pdf.

is approximately 7251 million kg. Rice production fluctuates over time, a feature attributed to market considerations (**Table 2**).

The temperature and rainfall observations are for Memphis, TN., given that this city is relatively equidistant from the northern and southern reaches of the Mid-South rice belt. The mean maximum temperature occurs in mid-July and averages 28°C (82°F) (**Figure 1**). The mean annual temperature is 17.5°C (63°F). The mean maximum precipitation occurs in April and averages 15 cm (6 inches). The driest month is September with 7.5 cm (3 inches). The mean annual total rainfall is near 140 cm (55 inches) (**Figure 2**). Considerable variation may occur, frequently attributed to tropical cyclones [2].

3. Typical rice growing practices in the Mid-South USA

Abundant radiant energy (17–28 MJ m^{-2} for July clear skies), land-graded soils that limit water percolation, plentiful groundwater for irrigation permit profitable rice production. Mid-South rice production is characterized as drill-seeded and delayed flood with ponding initiated at the 4th or 5th leaf stage. During dry periods, fields are "flushed" with a temporary and shallow flood to promote seed germination and herbicide activation.

Two common nitrogen fertilization programs include: (i) a two-way or three-way split with approximately 65–75% of the total nitrogen applied at flood, with the remainder applied in one or two airplane applications starting at internode elongation, or (ii) all nitrogen applied preflood. Recently furrow irrigated rice producers have been experimenting with additional nitrogen applications prior to internode elongation [3]. Urea (46-0-0) and ammonium sulfate (21-0-0) 24% S are commonly used nitrogen fertilizers. Rice-soybean (*Glycine max* (L.) Merr.) and continuous rice are common rotations. In Louisiana and Texas, producers frequently employ a ratoon rice crop, given their longer growing season. Field draining and soil drying provides for mechanical harvest. On-farm rice storage is a common practice.

Recently, rice producers have adopted furrow irrigation on graded-land [3]. In furrow irrigated rice, groundwater is conveyed in flexible plastic tubing and applied as side inlet water. Producer advantages of furrow irrigation include: (i) water conservation, (ii) reduced levee construction, (iii) lower production costs, and (iv) smaller arsenic concentration in rice. Current producer disadvantages of furrow irrigation include: (i) greater emphasis on weed management, (ii) nitrogen losses attributed to nitrification-denitrification, (iii) lack of crop insurance, (iv) less predictable yield potential, (v) need for disease resistance in plant materials, and (vi) producer education [4].

4. Defining greenhouse gases, global warming potential and carbon dioxide equivalent

Greenhouse gases are trace atmospheric gases that absorb outgoing long wavelength electromagnetic radiation, thus increasing the atmospheric temperature. Water vapor is a greenhouse gas; however, water vapor is treated separately from other greenhouse gases. Global warming potential is an estimate of the energy absorbed from a unit mass of a specific gas compared to carbon dioxide's energy absorption. Generally, the amount of energy absorbed is estimated for a 100-year time frame. Global warming potential values for methane vary from 25 to 36, with an atmosphere residence of 10 years [5–7]. Nitrous oxide has global warming potential values ranging from 265 to 298, with an atmosphere residence time of more than 100 years. Carbon dioxide accounts for approximately 75% of the global greenhouse gas emissions, an attribute primarily associated with fossil fuel combustion, deforestation, and biomass decomposition. Carbon dioxide concentrations during the Industrial Revolution were approximately 280 ppm, current CO_2 concentrations are greater than 395 ppm CO_2 [7].

5. Influence of temperature increases on rice growth and development

The specter of increasing temperature associated with climate change may impact rice production [8]. The prevalent modern view is that high temperature-induced spikelet sterility is a serious concern. Many studies have focused on high nighttime temperatures as more critical than high daytime temperatures. In a review, Wichelns [9] predicted that increased minimum and maximum daily temperatures and changes in the timing, intensity, and duration of rainfall will occur and increased temperatures may intensify the likelihood of spikelet sterility. High temperature impacts on pollen viability and spikelet sterility are more prominent in indica cultivars than japonica cultivars [6]. Ali et al. [10] noted that rice plants are particularly sensitive to high temperature stress during microspore and megaspore formation and that increased spikelet sterility will reduce yield. Rice yield losses because of applied heat stress were greatest from panicle exertion to anthesis; however, there were significant genotype and heat stress interaction differences. Thus, rice breeding efforts to provide heat tolerant cultivars is warranted.

6. Influence of irrigation on rice yield and quality

Ye et al. [11] asserted that long-duration varieties better tolerated the impact of climate change. In California. Lui et al. [12] completed a soil fertility study involving nitrogen, phosphorus, potassium, and silicon with contrasting high daytime temperatures. High day time temperatures decreased grain yield and decreased the transference of nitrogen, phosphorus, and potassium to panicles. Under conditions of high day time temperatures, silicon fertilization improved grain yield and the nitrogen, phosphorus, and potassium translocation to panicles.

Climate change may support CO_2 enrichment and yield advancement; however, grain quality may be negatively affected. In Arkansas, Esquerra et al. [13] tested rice varieties at (i) flag leaf collar and (ii) initial elongation of grain on the panicle for two elevated nighttime temperatures. The control temperature was 23°C and a high night-time temperature was 28°C. Spikelet fertility and yield were reduced for some varieties for high nighttime temperatures imposed at flag leaf collar.

Based on a compelling review, Horie [6] reiterated that given an increase in the CO_2 atmosphere content that the literature results include: (i) under optimum nitrogen availability rice biomass increased 24% between 24 and 31°C, (ii) panicle weight reductions at temperatures greater than 29°C were attributed to spikelet sterility, (iii) leaf area index was consistent regardless of temperature and the increased biomass was attributed to enhance photosynthesis, (iv) water use efficiency was improved because of the increase of biomass and photosynthesis, (v) panicle yield increased because of improved tillering, (vi) indica genotypes showed higher yield responses to enhanced atmosphere CO_2 than Japonica genotypes, and (vii) spikelet sterility was attributed to pollination failure.

Arnell et al. [14] investigated global and regional impacts because of increased temperatures arising from increased major heat wave frequencies, hydrological droughts, reduction in crop growth durations, increased river flooding, and others. Xie et al. [15] estimated crop yields and planted acreages along the Lower Mississippi River and inferred that US Mid-South rice yields would slightly decrease; however, planted acreages were estimated to increase because of land expansion to rice at the expense of corn (*Zea mays* L).

7. Sustainability of ground water resources

In Missouri, as in other rice producing states, groundwater resources substantially augment rainfall for rice and other row crops. The Mississippi River Valley Alluvial Aquifer in Arkansas, Mississippi, and Missouri is the major ground water resource for rice irrigation. In general, the Mississippi River Valley Alluvial Aquifer is typically an unconfined aquifer in Missouri, whereas the aquifer varies from an unconfined to confined aquifer in Arkansas [16]. Aquifer recharge is by rainfall and base flow from the Mississippi River, other rivers and streams, drainage ditches, and surface water bodies. Critical Ground-Water Areas are aquifers where water declines of 0.3 m (1 ft) occur for a minimum of five consecutive years. The Grand Prairie Region of Arkansas, composed primarily of Jefferson, Arkansas, Lonoke, and Prairie Counties, has been certified as a critical ground-water area, which predisposes producers to more detailed extension services to protect water resources. In Louisiana, excessive groundwater withdrawal from the Coastal Lowland Aquifer system is also a concern.

The Mississippi River Valley Alluvial aquifer is largely composed the sands and gravels (valley train), which are overlain by sandy to clayey alluvium. The confining unit, where present, is typically composed of silty and clayey alluvium in floodplains, backswamp, and meanderbelt sediments. In Missouri the confining unit is approximately 6–9 m (20–30 ft), whereas the confining unit in west-central Mississippi may be greater than 30 m (100 ft) and in the Grand Prairie Region of east-central Arkansas is greater than 18 m (60 ft). The Mississippi River Valley Alluvial aquifer is incised by multiple rivers, including the Mississippi River, White River, Cache River, St. Francois, and St. Charles River. Natural aquifer recharge may occur by (i) rainfall, (ii) upward flow from deeper aquifers, and stream aquifer flow (base flow) [16, 17].

There is growing concern that ground water withdrawal rates exceed recharge rates, leading to declining water tables or deeper equipotential surfaces [18]. Currently in Arkansas significant cones of depression exist in the Grand Prairie region and west of Crowley's Ridge [18]. In Missouri, Arkansas, and Mississippi the Mississippi River Valley Alluvial aquifer water level data strongly indicates that groundwater utilization is exceeding replenishment rates, indicating the water utilization rates are not sustainable. Water level declines are especially apparent in regions where water utilization rates are correspondingly high, especially in the Grand Prairie and Cache River areas. There is some correlation between the intensity of the groundwater drawdowns and their distance from major water streams and rivers, especially the Mississippi, Arkansas, and White Rivers.

The underlying Mississippi Embayment Aquifer has an extensive areal extent, ranging from Missouri to the coastal lowland aquifers. The hydraulic conductivity is substantial, ranging from 61 to 75 m d^{-1} (200 to 245 ft. d^{-1}) in eastern Arkansas and northeastern Louisiana to more than 75 m (245 ft. d^{-1}) is southeastern Missouri.

The underlying Mississippi Embayment Aquifer is partitioned into the Claiborne and Wilcox aquifers and the deeper McNary-Nacatoch aquifer [16, 17].

The United States Geological Survey (USGS) has established a series of observation wells across the USA [https://www.usgs.gov/mission-areas/water-resources/data-tools] (verified April 2023) [19]. Selecting one observation well from Missouri illustrates regional differences in aquifer overdraft potential (**Figure 3**). Located in Qulin, Missouri, the observation well shows that the water table drawdowns only exist during the irrigation season, after which the water table levels return to their pre-irrigation levels. Rainfall and base flow return consistently refurbishes the aquifer (**Figure 4**).

Table 3 provides examples of 16 US Geological Survey monitoring wells from Missouri, Arkansas, and Mississippi. In Missouri, every monitoring well indicates recharge every year, whereas Arkansas demonstrates some wells having consistent annual recharges and other wells exhibiting decades-long declines.

Water level declines in the Mississippi River Valley Alluvial Aquifer are attributed largely to withdrawals for rice (*Oryza sativa* L.) irrigation [20]. For Arkansas the 2020 estimate of groundwater application in agriculture was 25,380,820 m^3 d^{-1} (5583 Mgal d^{-1}) from the Mississippi River Valley Alluvial aquifer and the deeper Sparta aquifer. The withdraw from the Mississippi River Valley Alluvial aquifer was 23,148,690 m^3 d^{-1} (5092 Mgal d^{-1}). The sustainable yield is estimated as 15,338,508 m^3 d^{-1} (3374 Mgal d^{-1}) [17]. For a considerable time-frame, the Big Sunflower River and lower Yazoo River have drained the alluvial aquifer [17]. Massey et al. [21] noted that the US mid-South region contains four million ha irrigated croplands. When averaged across all years and irrigation methods, irrigation rates were 9200 m^3 ha^{-1} for rice, substantially greater than for corn (*Zea mays*), soybean (*Glycine max*), and cotton (*Gossypium hirsutum*).

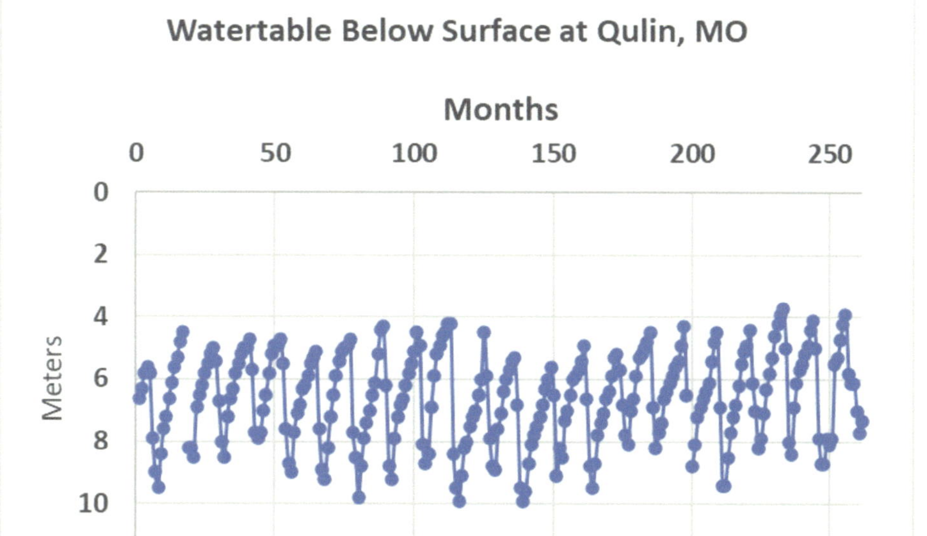

Figure 3.
Depth to water table (meters) from United States geological survey well in Qulin, Missouri. Data from USGS. https://waterdata.usgs.gov/monitoring-location/363551090152801/#parameterCode=72019&period=P7D.

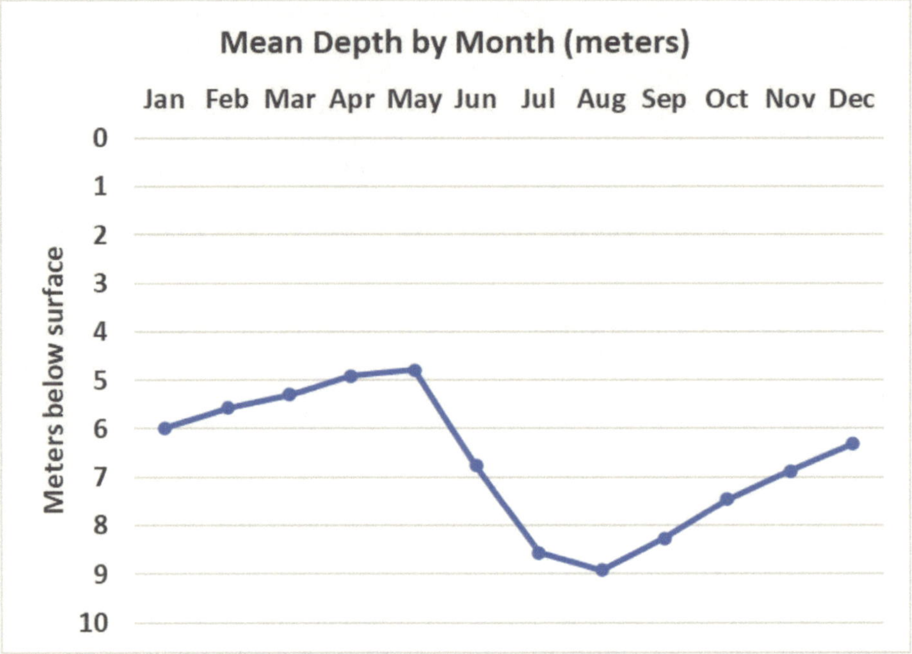

Figure 4.
Mean monthly depth to water table (meters) for Qulin, Missouri. Data from USGS. https://waterdata.usgs.gov/monitoring-location/363551090152801/#parameterCode=72019&period=P7D.

Location	Depth (m)	Range (m)	Aquifer recharge estimate
Steele, MO	3.3	1.5–5.3	Recharge every year
East Prairie, MO	2.4	1.2–3.7	Recharge every year
Cardwell, MO	3.0	0.9–4.3	Recharge every year
Sikeston MO	2.1	1.0–2.7	Recharge every year
Naylor, MO	2.1	1.2–4.3	Recharge every year
Caruthersville, MO	5.8	5.7–6.8	Recharge every year
Jackson Co. AR	—	16.8–22.3	Decline for 32 years
Phillips Co, AR	4.3	1.5–6.0	Recharge every year
Drew Co, AR	—	6.2–10.4	Decline for 60+ year
Jonesboro, AR	—	17–9.2	Decline for 55+ year
Crittenden Co, AR	7.6	7.3–8.2	Recharge every year
Mississippi Co, AR	3.7	1.5–4.3	Recharge every year
Arkansas Co, AR	—	29.3–31.1	Slight decline for 63+ year
Cross Co, AR	—	26.8–26.5	Very slight decline for 19 year
Craighead Co, AR	25.9	24.4–26.8	Recharge every year
Sunflower Co, MS	7.9	3.6–9.4	Slight decline for 12+ year

Source: United States Geologic Survey [19].

Table 3.
Examples of Groundwater depths and assessment of aquifer recharge.

8. Climate change prospects for the Mid-South USA and technology-based mitigation strategies

It is likely that the US Mid-South region will witness (i) greater likelihood of inland water flooding because of increasing rainstorm intensities, especially in fall and winter, (ii) drought risk, and (iii) greater frequency of daytime temperatures exceeding 35°C (95°F). These forecasts imply that (i) flood-induced delayed spring plantings may be more frequent, (ii) increased irrigation and greater likelihood of aquifer overdraft, (iii) need for high-temperature tolerant cultivars, (iv) amplified insect, disease and weed pressures imply the creation of improved integrated pest management approaches, and (v) augmented producer-oriented education programs promote emerging technology and farmgate acceptance [22].

Reviewing the literature, Walthall et al. [22] noted that wheat (*Triticum aestivum*), soybeans (*Glycine max*) and rice are likely to experience 12–15% yield reductions upon transition from atmospheric CO_2 concentrations from 370 ppm to 550 ppm. Walthall et al. also presented evidence suggesting that temperatures increase of 1–2°C are expected. Aggregate affects of climate change include: (i) soil behavior, (ii) soil erosion, (iii) wind and humidity changes, (iv) insect and disease incidents, (v) weed growth characteristics, and (vi) invasive organisms.

One important and potentially transformational climate change mitigation practice focuses on irrigation water efficiency. The introduction of (i) multiple-inlet irrigation, (ii) tailwater recovery, (iii) surge pumps permit uniform depth of soil water infiltration. Recently alternate wetting-drying and furrow irrigation strategies have the potential to reduce field runoff, employ smaller amounts of water and maintain high rice yields [23].

9. Intermittent flood or alternate wetting and drying

Intermittent flood or alternate wetting and drying was initially developed at the International Rice Research Institute (Philippines), where at a minimum of three weeks after the initial flood the flood is permitted to recede to the point where the soil surface is either wet or dry, then reflooding occurs [24]. Flood should be present at panicle initiation, anthesis, and grain fill. The reflood scenarios are influenced by growth stage, irrigation capacity, and risk tolerance. Advantages of alternate wetting and drying include greater capacity for rainfall capture, reduced methane emission, reduced arsenic accumulation, and potential likelihood of groundwater conservation [24].

Atwill et al. [25] compared conventional flood, multiple side inlet with and without alternating wetting and drying. Multiple side inlet with alternating wetting and drying required 39% water applications and rough rice yield was not significantly different across irrigation regimes. Irrigation water use efficiency was 59% greater for alternate wetting and drying. Where groundwater levels were reasonably deep, the net production returns were more positive because of the reduced costs of water pumping. In southeastern China, Yang et al. [26] compared controlled irrigation and traditional flood irrigation. A decrease of 46% irrigation water usage did not influence rice yield but did improve water use efficiency. In Arkansas, Chlapecka et al. [27] compared furrow irrigation and alternate wetting and drying rice systems, demonstrating that (i) the alternate wetting and drying system favored water conservation and (ii) rice yields were comparable, supporting a greater water use efficiency. In Missouri, multiple years of furrow-irrigation of rice shows promise in maintaining yields; however, substantial issues remain in securing a consistent (i) nitrogen fertilization regime, (ii) weed management program, and (iii) irrigation timing protocol [4, 28, 29]. In a greenhouse project, Lunga et al. [30] documented that the above ground biomass was greatest for the flood system.

The International Rice Research Institute listed potential advantages for alternate wetting and drying [24]. With continuing research, some of the alternate wetting and drying advantages included: (i) potential for cultivars with improved characteristics for aerobic production, (ii) improved ecosystem services, human health, and environmental protection, (iii) improved soil fertility, weed management, and integrated pest management practices, and (iv) reduced soil sodium accumulation because of elevated sodium adsorption ratios.

In India, Shekhar et al. [31] conducted alternate wetting and drying field research involving three levels of soil moisture depletion with sub-treatments involving nitrogen management. Yields for conventional irrigation with alternate wetting and drying imposing only mild stress were statistically equivalent. Alternate wetting and drying imposing severe moisture stress resulted in a 9% rice yield reduction. The nitrogen use efficiency was similar for conventional irrigation and mild stress alternate wetting and drying. Atwill et al. [25] performed field research in Louisiana and Mississippi comparing conventional irrigation and alternate wetting and drying and documented that rice grain yield for six cultivars was statistically equivalent.

Carrijo et al. [32] conducted a meta-analysis to identify soil properties and land management that favored successful alternate wetting and drying and estimate the water conservation potential and differences in rice yield. The conventional flood and mild stress alternate wetting and drying irrigation regimes showed similar rice yields and the mild stress irrigation regimes demonstrated small water usages. Carrijo et al. [33] imposed three alternate wetting and drying irrigation treatments having increasing water severity between full canopy cover and 50% heading. The experimental data suggests that the availability of soil water in the 25–35 cm soil rooting depth was critical for maintaining rice yields.

10. Furrow irrigated rice

In Arkansas, Frizzell et al. [34] compared twelve cultivars in a furrow irrigation trial, where rice yields were evaluated at the side-inlet portion of the field with the bottom portion of the field where tailwater accumulated. Rice yields and headrice yields were greater in the bottom portion of the field, especially for the hybrids. In Arkansas, Henry and Clark [35] evaluated furrow irrigated rice having six different irrigation timings, that is, continuous, 3, 5, 7, 10 and 14 days. The continuous water timing had the highest yield and the poorest water use efficiency, whereas the 14-day irrigation frequency had the lowest yield and the highest water use efficiency. Irrigation having a 40% allowable depletion was considered the threshold for yield penalties.

In a two-year field study in Missouri, Aide and Goldschmidt [4] compared 12 and 17 varieties for their yield and arsenic uptake because of furrow irrigation or delayed flood irrigation. The plots were 450 m long and divided into three sections based on the distance from the side-inlet water application. The tail water accumulation portion of the field and for the delayed flood irrigation exhibited the greatest yields and the highest arsenic levels in straw, rough rice, and milled rice. The plot portion adjacent to the side-inlet irrigation exhibited the lowest yields and the smallest arsenic levels. Similar results were documented for the same experimental design in subsequent years [28, 29].

11. Projections of needed rice research to address climate change and unsustainable water use

With the threat of climate change, the mid-South rice region is anticipating greater temperatures and uncertainty in precipitation. Additionally, the irrigation water application rates is challenging aquifer sustainability. These two issues collectively influence rice production and quality. With nighttime temperatures approaching 29°C (84°F) the likelihood of spikelet sterility becomes an increasing threat to production. Cultivars do show a range of heat stress, thus favorable results of dedicated rice breeding programs are quite possible. Aquifer overdraft is recognized on a regional basis, which reduces the long-term sustainability of rice production. Producer adoption of either furrow irrigated rice or alternate wetting and drying irrigation has substantial potential to reduce aquifer overdraft. Thus, two technological solutions may mitigate the deleterious influence of climate change. Other water conserving solutions include: (i) surface water capture, (ii) tailwater recovery, and (iii) development of constructed wetlands.

12. Conclusion

The specter of climate change is real and the effects on rice production will certainly develop. Conversely, advances in technology may reduce rice yields and impair quality degradation; however, substantial refinement of these technologies must be implemented before there is widespread commercialization. The development of rice plant genetics to improve high temperature tolerant cultivars and the continuing development and producer adoption of water conserving irrigation regimes are critical to mitigating the influence of climate change. Other agronomic concepts are potentially viable; however, they will not be sufficient without improved cultivars and innovative irrigation practices. Furrow irrigation is becoming more accepted and considerable rice acreage is actively using this irrigation regime. Research that supports consistent yields and rice quality will increase producer implementation and subsequently will support aquifer sustainability.

Conflicts of interest

The author declares that there is no conflict of interest.

Author details

Michael Aide
Department of Agriculture, Southeast Missouri State University,
Cape Girardeau, MO, USA

*Address all correspondence to: mtaide@semo.edu

References

[1] United States Department Agriculture-National Agricultural Statistics Service. Available from: https://www.nass.usda.gov [Accessed: April 2023]

[2] National Weather Service. Available from: https://www.weather.gov/media/meg/August2017ClimateSummary.pdf [Accessed: April 2023]

[3] Nalley LL, Linquist B, Kovacs KF, Anders MM. The economic viability of alternate wetting and drying irrigation in Arkansas rice production. Agronomy Journal. 2015;**107**:579-587

[4] Aide MT, Goldschmidt N. Comparison of delayed flood and furrow irrigation involving rice for nutrient and arsenic uptake. International Journal of Applied Agricultural Research. 2017;**12**:129-136. Available from: https://www.ripublication.com/ijaar.htm

[5] Simmonds MB, Anders M, Adviento-Borbe MA, van Kessel C, McClung A, Linquist BA. Seasonal methane and nitrous oxide emissions of several rice cultivars in direct-seeded systems. Journal of Environmental Quality. 2015;**44**:103-114

[6] Horie T. Global warming and rice production in Asia: Modeling, impact prediction and adaptation. Proceedings of the Japan Academy Series B: Physical and Biological Sciences. 2019;**95**(6):211-245. DOI: 10.2183/pjab.95.016

[7] Conrad R. The global methane cycle: Recent advances in understanding the microbial processes involved. Environmental Microbiology Reports. 2009;**1**:285-292. DOI: 10.1111/j.1758-2229.2009.00038x

[8] Krishnan P, Ramakrishnan B, Reddy KR, Reddy VR. High-temperature effects on rice growth, yield, and grain quality. Advances in Agronomy. 2011;**111**:87-206

[9] Wichelns D. Managing water and soils to achieve adaptation and reduce methane emissions and arsenic contamination in Asian rice production. Water. 2016;**8**:141-170. DOI: 10.3390/w8040141

[10] Ali MA, Hoque A, Kim PJ. Mitigating global warming potentials of methane and nitrous oxide gases from rice paddies under different irrigation regimes. Ambio. 2013;**42**:357-368. DOI: 10.1007/s13280-012-0349-3

[11] Ye T, Zong S, Kleidon A, Yuan W, Wang Y, Shi P. Impacts of climate warming, cultivar shifts, and phenological dates on rice growth period length in China after correction for seasonal shift effects. Climate Change. 2019;**155**:127-143. DOI: 10.1007/s10584-019-02450-s

[12] Liu Q, Ma H, Sun Z, Lin X, Zhou X. Translocation efficiencies and allocation of nitrogen, phosphorus and potassium in rice as affected by silicon fertilizer under high daytime temperature. Journal of Agronomy and Crop Science. 2019;**205**:188-201. DOI: 10.1111/jac.12313

[13] Esguerra MQ, Hemphill CC, Counce PA. Differential response of arkansas rice varieties on high nighttime temperature (HNT) treatments at different reproductive stages. In: KAK M, Scott B, Hardke B, editors. BR Wells Arkansas Rice Research Studies 2019. Arkansas Agricultural Experiment Station, Research Series 667. Fayetteville, Arkansas: University of Arkansas ystem, Division of Agriculture; 2019

[14] Arnell NW, Lowe JA, Challinor AJ, Osborn TJ. Global and regional impacts of climate change at different levels of global temperature increase. Climate Change. 2019;**155**:377-391. DOI: 10.1007/s10584-019-02464-z

[15] Xie L, Lewis SM, Auffhammer M, Berck P. Heat in the heartland: Crop yield and coverage response to climate change along the Mississippi River. Environmental Resource Economics. 2019;**73**:485-513. DOI: 10.1007/s10640-018-0271-7

[16] United States Geological Survey. Ground Water Atlas of the United States. 2016. Available from: https://pubs.usgs.gov/ha/ha730/gwa.html [Accessed: May 2023]

[17] Arkansas Department of Agriculture. 2022 Arkansas Groundwater Protection and Management Report. 10421 West Markham Street, Little Rock, AR: Arkansas Department of Agriculture – Natural Resources Division; 2022

[18] Reba ML, Massey JH, Adviento-Borbe MA, Leslie D, Yaeger MA, Anders M, et al. Aquifer depletion in the lower Mississippi River basin: Challengers and solutions. Journal of Contemporary Water Research and Education. 2017;**162**:128-139

[19] United States Geologic Survey. Available from: https://www.usgs.gov/mission-areas/water-resources/data-tools [Accessed: April 2023]

[20] Atwill RL, Krutz LJ, Bond JA, Golden BR, Spencer GD, Bryant CJ, et al. Alternate wetting and drying reduces aquifer withdrawal in Mississippi rice production systems. Agronomy Journal. 2020;**112**:5115-5124. DOI: 10.1002/agj2.20447

[21] Massey JH, Mark SC, Epting JW, Shane PR, Kelly DB, Bowling TH, et al. Long-term measurements of agronomic crop irrigation made in the Mississippi delta portion of the lower Mississippi River valley. Irrigation Science. 2017;**35**:297-313

[22] Walthall CL, Hatfield J, Backlund P, Lengnick L, Marshall E, Walsh M, et al. Climate Change and Agriculture in the United States: Effects and Adaptation. Washington, DC: USDA Technical Bulletin 1935; 2012. 186 p. Available from: https://www.usda.gov/sites/default/files/documents/CC%20and%20Agriculture%20Report%20(02-04-2013)b.pdf

[23] Aide MT. Rice production with restricted water usage: A global perspective. Egyptian Journal of Agronomy. 2019;**41**:197-206. DOI: 10.21608/agro.2019.15729.1174

[24] Allen JM, Sander BO. The Diverse Benefits of Alternate Wetting and Drying (AWD). Los Banos, Philippines: International Rice Research Institute; 2019. Available from: www.ccafs.cgiar.org

[25] Atwill RL, Krutz LJ, Bond JA, Reddy KR, Gore J, Walker TW, et al. Water management strategies and their effects on rice grain yield and nitrogen use efficiency. Journal of Soil and Water Conservation. 2018;**73**:257-264. DOI: 10.2489/jswc.73.3.257

[26] Yang S, Liu X, Liu X, Xu J. Effect of water management on soil respiration and NEE of paddy fields in Southeast China. Paddy and Water Environment. 2017;**15**:787-796. DOI: 10.1007/s10333-017-0591-1

[27] Chlapecka JL, Hardke JT, Roberts TL, Frizzell DL, Castaneda-Gonzalez E, Clayton T, et al. Allowable water deficit when utilizing alternative rice irrigation strategies. In: KAK M, Scott B, Hardke B, editors. BR Wells Arkansas Rice Research

Studies, Arkansas Agricultural Experiment Station, Research Series. Vol. 667. Fayetteville, Arkansas: University of Arkansas System, Division of Agriculture; 2019

[28] Aide MT. Comparison of delayed flood and furrow irrigation regimes in rice to reduce arsenic accumulation. International Journal of Applied Agricultural Research. 2018;**13**:1-8. Available from: https://www.ripublication.com/ijaar.htm

[29] Aide MT. Furrow irrigated rice evaluation: Nutrient and arsenic uptake and partitioning. International Journal of Applied Agricultural Research. 2019;**14**:15-21. ISSN 0973-2683

[30] Lunga D, Brye KR, Slayden JM, Lebeau SG, Roberts TL, Norman RL. Water management effects on trace gas emissions under greenhouse conditions from direct-seeded hybrid rice in a silt-loam soil. In: KAK M, Scott B, Hardke B, editors. BR Wells Arkansas Rice Research Studies, Arkansas Agricultural Experiment Station, Research Series. Vol. 667. Fayetteville, Arkansas: University of Arkansas System, Division of Agriculture; 2019

[31] Shekhar S, Mailapalli DR, Raghuwanshi NS. Effect of alternate wetting and drying irrigation practice on rice growth and yield: A lysimeter study. ACS Agricultural Science & Technology. 2022;**2**(5):919-931. DOI: 10.1021/acsagscitech.1c00239

[32] Carrijo DR, Lundy ME, Linquist BA. Rice yields and water use under alternate wetting and drying irrigation: A meta-analysis. Field Crops Research. 2016;**203**:173-180. DOI: 10.1016/j.fcr.2016.12.002

[33] Carrijo DR, Akbar N, Reis AFB, Li C, Gaudin ACM, Parikh SJ, et al. Impacts of variable soil drying in alternate wetting and drying rice systems on yields, grain arsenic concentration and soil moisture dynamics. Field Crops Research. 2018;**222**:101-110. DOI: 10.1016/j.fcr.2018.02.026

[34] Frizzell DL, Hardke JT, Amos LR, Castaneda-Gonzalez E, Clayton TL. Performance of twelve rice cultivars in a furrow-irrigated rice (FIR) system. In: Hardke J, Sha X, Bateman N, editors. B.R. Wells Arkansas Rice Research Studies. Arkansas Agriculture Research Series, University Arkansas; 2021. p. 231

[35] Henry CG, Clark T. Evaluating irrigation timing, depletion, water-use, and efficiencies in furrow-irrigated rice. In: Hardke J, Sha X, Bateman N, editors. B.R. Wells Arkansas Rice Research Studies. Arkansas Agriculture Research Series, University Arkansas; 2021. p. 244

Chapter 3

Enhanced Agricultural Productivity Using Hydroponics Technique: A Smart Farming System

Suman Dutta, Bishal Mukherjee and Ashutosh Sawarkar

Abstract

Hydroponic farming is one potential solution to the lack of arable land diminishing the capacity of agriculture. The hydroponic method of crop production has proved successful for precision farming in growing both flowers and vegetables. It requires fewer energy requirements than traditional agriculture because it employs fertilizer solutions under heavily controlled environmental conditions in limited areas. Hydroponic systems can be used as a treatment method for partially treated wastewater or reclaimed water before its discharge into the environment since plants have the ability to absorb nutrients, toxic metals, and emerging contaminants. Farmers engaged in hydroponic farming benefit from a wide range of significant advantages by enhancing their income through introducing quality products for a sustainable community. The newly created technology also arrived at the perfect time because traditional farming practices do not work with diminishing water levels. Plants may now be grown in any greenhouse or nursery, regardless of the season, as long as the necessary infrastructure is in place.

Keywords: hydroponics, precision agriculture, greenhouse facility, enhanced farm income, soil less cultivation

1. Introduction

Environmentally friendly soilless agricultural production methods control inputs like water, pesticides, and fertilizers to produce farm output. Furthermore, soilless agriculture makes the best use of available energy. It offers significantly cleaner and hygienic products than soil-based agriculture [1]. The technology required to use soilless agricultural techniques has rapidly improved, expanded, and gained a lot of significance in recent years. This technology is relevant to farming methods that use minimal water and little land while allowing total control over all agricultural variables. A number of things affect traditional farming [1]. A number of elements, including soil nutrients, climatic circumstances, and inappropriate soil structures for cultivation, may have a negative effect on conventional agriculture.

Additionally, soil-borne pathogens and pests can dramatically lower overall productivity. Consequently, the use of agricultural methods in a controlled environment has become increasingly important recently. The most significant kind is "soilless agriculture," which has gained popularity as a result of changing climatic conditions and dwindling agricultural land [2]. Approximately 12% of the earth's surface (1.5 billion hectares) is used for agriculture [3]. According to the FAO's 2030/2050 forecast assessment, the amount of arable land per person in developing countries, wealthy countries, and the entire world would decrease annually. Although both the world population and food consumption are expanding, the average quantity of arable land per person is decreasing yearly [3]. This suggests that some countries may face food crises in the future. Additionally, the amount of arable land is decreasing daily due to climate change and the misuse of agricultural land for non-agricultural purposes [3]. Nations have looked for a variety of solutions to deal with the population expansion that threatens food security. The solution currently requires international agricultural investment and soilless farming methods. The depletion and contamination of freshwater resources are additional key problems in agriculture, in addition to the decreasing area of arable land.

One of the well-liked methods of indoor soilless farming is hydroponics, which uses less fertilizer and offers greater protection against pests and adverse weather conditions [4]. High-value crops can be grown hydroponically better than low-value field crops [5]. In open fields and naturally ventilated poly buildings at farming systems, a variety of cool-season vegetables were assessed utilizing vegetative and yield metrics [6]. Using conventional agronomical approaches, the yield of a crop is significantly impacted by climatic changes, pests, and diseases, which results in low-quality output. Using the hydroponic technique, mineral fertilizer solutions dissolved in water act as a solvent to enhance plant uptake of soil nutrients. The hydroponics techniques are divided into groups based on the availability, positioning, and method of administering nutrient solutions to plant roots [6]. It has been discovered that crops produced in hydroponic systems produce more continually throughout the year and need less time to grow than crops cultivated in conventional systems. The growing medium in hydroponics directs water and nutrients while enabling ample oxygen to reach the plant's roots. Different types of hydroponic systems work well with varied growing materials [7]. Despite the absence of soil-borne pests and diseases, soilless growth techniques like hydroponic systems have some risks, most notably the existence of waterborne pathogens [8]. These risks negatively affect hydroponic recirculating systems where bacteria can build up over time. The issues with hydroponics also originate from the requirement for financial outlays and technical expertise to use the control systems. Rufi-Salis et al. [9] explore the many hydroponic fertilizer usage reduction strategies in order to stop nutrient discharge into the environment. Closed systems that use water efficiently and preserve space have quickly become vital under these circumstances when health problems and access to wholesome food are significant difficulties. Even those with no interest in farming have recognized the value of organic farming and attempted to develop its products.

2. Different hydroponic cultivation techniques

Contrary to conventional farming, hydroponics does not use soil to grow food. This method involves growing plants on artificial or natural substrates so that the roots may easily draw nutrients from a prepared nutrient solution (**Figure 1**).

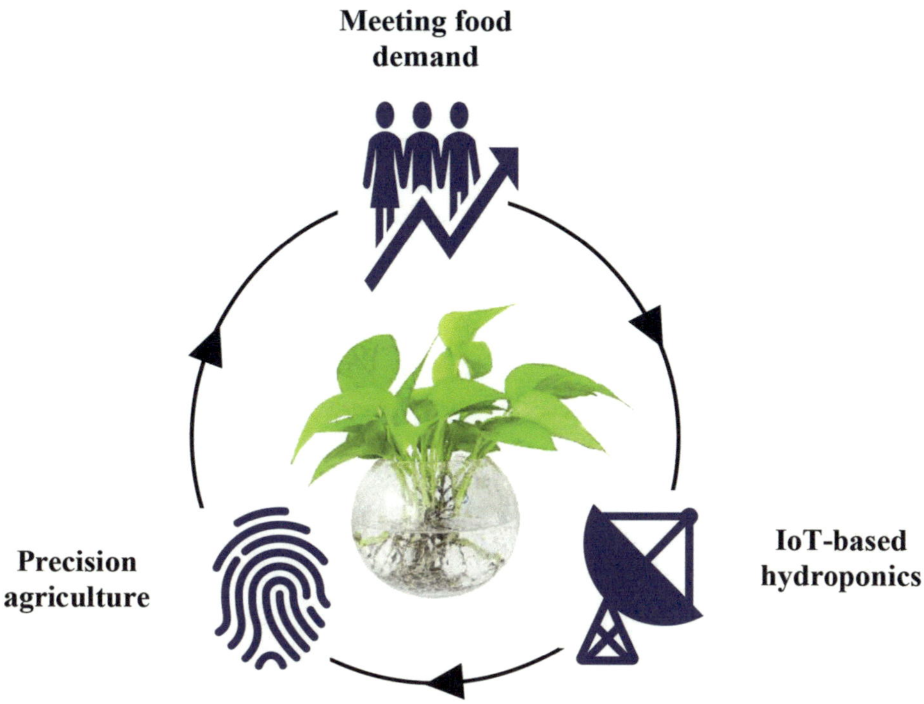

Figure 1.
Interconnection of hydroponic with precision agriculture and internet of things to meet global food demand.

The implementation of various hydroponic farming techniques varies depending on the type of plant, regional climate, and financial constraints, among other things [10]. In Floating Root System or Deep Water Culture (DWC) system, plant is held above the water line by polystyrene, cork bark, or wood, among other materials, with only the root submerged in the nutrient solution. In drip irrigation, a controlled flow of fertilizer solution is injected straight into the plant roots. At regular intervals, the solution is administered, and in closed systems, any remaining solution is disposed of in the storage tank [10]. In aeroponics, a sprinkler system periodically sprays nutrients onto the plants, which have roots that hang down in the air. The key benefit of this method is that it does not need an airing system because oxygen is carried in the fertilizer solution that is sprayed and is the best method for growing tubers and roots. Nutrient film techniques (NFT) are similar to those in a floating root system, but instead of being entirely submerged in the nutrient solution, they are instead suspended in a liquid stream that is passing via a pipe system [10]. NFT needs more energy and components to function, even though it uses less nutritional solution than the floating root system. The flow of nutritional solution can be continuous or irregular, and the excess solution returns to the storage tank via gravity. In the Ebb and Flow method, a tray with plants in it receives regular fills of nutrient-rich water blasted up from a reservoir below. The technology employs gravity to recycle the water by returning it to the reservoir [10]. In contrast, the aquaponics method takes advantage of the symbiotic relationship between plants and animals to create a productive system where fish waste meets the nutritional needs of the plants. A healthy micro-ecosystem is created when fish tank water is recycled through the uptake of nutrients and the microbial nitrification and denitrification processes [10].

3. Essential tools required for indoor hydroponic system

Ten essential tools that indoor hydroponic gardeners should keep are a pH meter (to measure the acidic or basic nature of water), nutrient test kit (to measure different nutrients in the water), TDS meter (to measure dissolved solids, salt, minerals, and other concentrates in the water), timer (to track the watering time and lighting plants), thermometer (to keep track of the temperature in the grow room), hygrometer (helps you keep track of the humidity in the grow room), watering can or hose (to water plants regularly), scissor or trimmer (required for plant growth as it helps the plant focus its energy on producing more leaves, flowers, and fruits), measuring cups (used to measure the nutrients adding to the water), and grow room glasses (to protect your eyes from the bright grow lights and avoid any eye damage). The test equipment for automation kits used in soilless agricultural systems is expensive. As a result, the control kit acquired as part of the project will be quick and simple to use, enabling its widespread adoption due to its low cost and simplicity. A control kit was created using an Arduino microcontroller, four distinct sensors, add-on tools, and project coordinator-written software [1]. Although many industries use these controller cards now, their utilization in agriculture is not as high as it could be. Furthermore, such technology can regulate the temperature, pH, dissolved oxygen, electrical conductivity, and pH of the solution in systems that use the soilless farming method [1].

4. Nutrient solution used in hydroponics

With the exception of carbon, hydrogen, and oxygen, all necessary nutrients are delivered to the plant through the nutrient solution in hydroponics. With the exception of iron, which is given as an iron chelate to increase its availability, inorganic fertilizers are utilized as suppliers of nutrients. Although some inorganic acids are also employed, the majority of fertilizers used in hydroponics to generate nutrient solutions are highly soluble inorganic salts [11]. Depending on the crop and plant stage, the nutrient solution makeup varies. In hydroponics, soluble fertilizers including ammonium nitrate, calcium nitrate, phosphoric acid, and nitric acid are employed [11]. Although these formulations are sold commercially in liquid or solid form, the salt, mineral, and fertilizer mixture can also be made from scratch. It is crucial to note that different formulations are needed for different growth stages of plants. For instance, during the vegetative state, a plant grows foliage until it is ready to flower or ripen its roots, at which point the plant needs a nutrient solution high in phosphorus to develop strong roots. Finally, the plant needs nutritional solutions with high concentrations of K and low amounts of N during fruit ripening. According to the N-P-K concentration stated in weight percent, solutions that are commercially available codify the macronutrient contents as a three-digit sequence. Leaching the solution before adding it to the hydroponic system enables the addition of organic nutrients like compost, a mixture of vegetable waste, urine, manure, and dead animal parts. This kind of tea can potentially replace the conventional inorganic fertilizers used in hydroponics. However, adding such components to the formulation runs the risk of introducing unwanted parasites or bacteria, therefore it must be thoroughly examined before being added to the hydroponic system. If you want the solution to persist as long as possible, the pH, electrical conductivity, and water level must be adjusted as soon as feasible. In order to prevent variations in the nutrient solution, the volume level in the storage tank must be constant, replenishing the water received by the plants and lost through evapotranspiration. It was generally recommended to switch the tank solution every two to 3 weeks, depending on the crop, to provide the tank with a thorough cleaning and disinfection [12].

4.1 pH in hydroponics nutrient solutions

The pH of a nutrient solution, which is measured on a scale of 1–14 to indicate a solution acidity or alkalinity, is a crucial chemical characteristic. Water has a pH of 7, indicating that it is neither basic nor acidic at normal temperature. If the pH is more than 7, the solution is basic; otherwise, it is acidic. The majority of publications concur that the pH of the nutrient solution must be between 5 and 7, as nutrients stay soluble in this range [13]. The solubility of Fe and H2PO4 diminishes, causing Ca and Mg precipitates as well as other chemical interactions between the components of the nutritional solution, which prevents the absorption of iron, boron, copper, zinc, or manganese if pH is more than 7. However, the adsorption of nitrogen, phosphorus, potassium, calcium, magnesium, and molybdenum is prevented if pH is lower than 5. Hazardous contamination may occasionally be caused by the supply of some micronutrients, such as manganese [14].

4.2 Electrical conductivity of the solution

An estimation of the total ion concentration in a solution is electrical conductivity or EC. Low EC values in this situation point to a lack of nutrients in the form of ions, while high values could produce salt stress in the plant [15]. As a result, EC should be kept within a target range because it has a substantial impact on crop quality and growth [16]. Additionally, since this measure does not specifically indicate the concentration of each element in the nutrient solution, it is crucial to give fertilizers in concentrations that the plants can absorb after measuring EC.

4.3 Nutrient solutions sterilization

Hydroponic systems require an aseptic environment to effectively generate high-quality products, but it is challenging to maintain sterility in the area around plant roots [17]. The most noticeable sign of a damaged plant is leaf withering, which is brought on by the fungus Fusarium and Verticillium. The roots of the plant are also threatened by other parasite species including Pythium and Phytophthora. Unfortunately, using fungicides in hydroponics without endangering consumer health is not possible [18]. From a sustainability perspective, recirculating the nutrient solution lessens the quantity of water consumed and waste that must be disposed of, but it is not always practical to design systems that balance resource consumption, energy use, and cost. Given the advantages and disadvantages of each option individually, combining the hydroponic infection prevention techniques may be the best course of action to address the fertilizer solution sepsis problem.

5. Digital twin in smart agriculture

One effective soilless method is hydroponics, which has benefits including reduced water use, improved output, and the absence of chemical weed or pest control solutions. It is essential to use the most cutting-edge technologies available to increase hydroponics efficiency. IoT applications for smart hydroponics greenhouse farming are shown instrumental [19]. The primary benefit of a DT is that the virtual unit can forecast how its physical twin will function, thereby foreseeing potential problems and optimizing the entire system. Increasing output, reducing waste, protecting natural resources, and maintaining quality and other production standards are a few examples of this. Hydroponic and soil-based farms may respond differently to many environmental factors [20]. Predictive analysis and quick control are required in the hydroponics farming method for problems including nutrient solution management, pathogen management, weed management, and environment management. Among these, the dietary approach stands out as a crucial factor in determining high-quality production. The productivity and yield of the hydroponic farm are directly impacted by nutrient intake and pH control [21]. The availability of nutrients for plant uptake may be constrained by a higher pH value due to the presence of insoluble and inaccessible ions. The pressure that the ions of the nutritional solution exert on it can be measured using electrical conductivity (EC). The ideal EC for each crop varies based on the environment [22, 23]. Higher EC values translate into higher osmotic pressure, whereas lower EC values could be detrimental to the health and productivity of plants. However, it is not advisable to consistently apply nutritious solutions. According to studies, highly concentrated nutrient solutions encourage excessive nutrient consumption, which may have harmful effects [24]. Growers may benefit from an accurate predictive model on the right amount of nutrients to add for a certain crop at a specific time in order to reduce the negative effects of nutrient overuse. The capacity of hydroponic farms to adjust multiple parameters like pH, EC, and temperature can only be used with the aid of more accurate and predictive models. High nutrient solution temperature in hydroponics can cause a number of problems, including the plant refusing to absorb nutrients through the roots and finally going into survival mode [25]. This is similar to how EC and pH can affect the plant. Higher water temperatures in the fertilizer solution are also conducive to microorganisms rapidly proliferating. These may also develop in the root zone, which may cause several plant diseases [26]. Even the frigid temperatures of nutrient solutions are not ideal for plant growth. Throughout the course of a plant's life, the root zone should be a comfortable environment [6].

Digital twins (DT) encompass a variety of technologies, including artificial intelligence, augmented reality, the Internet of Things, communication technologies, embedded technologies, big data, processing techniques, data security, and cloud computing, among others [27]. DT can be characterized as an adaptive model of a living or non-living physical system that aims to construct, monitor, and enhance the performance of its actual counterpart and give end users a more realistic experience [28]. The lifecycle of a real machine, for example, can be applied to DT at any point in time. DT can be applied to the design of the product, to enhance the physical model, and to perform preventative maintenance. Despite the fact that DT technology actively participates in the development and marketing of several manufacturing and aviation. With the use of meteorological information, the DT idea can be used to improve the forecasting of nutrient solution temperature [29]. A predictive model for the temperature of the nutrient solution can be created by utilizing DT to determine the relationship between nutrient solution temperature and meteorological parameters. Farmers can utilize DTs to predict how well such cooling devices would function when they are used in a hydroponics farm without actually installing the actual units. It assists the farmers in developing a practical first plan for their farm and evaluating the performance impacts of adding new resources, including fans and heaters.

6. IoT-based hydroponics system

The application of modern information and communication technologies (ICTs) in agriculture is the foundation of the high-tech, expensive "smart farming" system. In IoT-based smart farming, a system is developed for sensor-assisted crop field monitoring and irrigation system automation [30]. Two businesses, ATLAS Scientific and Libelium Waspmote, offer electronic sensors and microcontroller-based systems for detecting environmental factors and water quality [31]. According to the surveys, many scientific articles have examples of these solutions [32, 33]. It is essential to use other important ICTs, such as big data [34], blockchain [35], and neural networks [36], which have been successfully employed in other scenarios like these to produce precise and efficient agriculture. Examples of IoT applications in agriculture include open-field farming and greenhouse cultivation [37, 38]. The most often observed metrics are temperature and humidity [37, 38], however other elements including light [39] and crop-specific traits [37] are also regularly noted. Several studies monitored variables such as soil moisture, temperature, humidity, and light in order to automate the irrigation system [39, 40]. The iPONICS system makes an effort to take advantage of some of the technologies mentioned above in order to offer a simplified, inexpensive, yet creative hydroponic solution that can be used by even hobby-type hydroponic systems [31]. The design and execution of a cutting-edge, low-cost IoT-based hydroponics monitoring and control system were demonstrated in many circumstances [31]. The system is made up in particular of a specialized wireless sensor network for controlling the watering process and monitoring the crucial hydroponics parameters. It offers the customer a simple web-based application to keep tabs on their crops while also alerting them with the proper alarms and warnings. This makes it much easier to observe numerous hydroponic greenhouses with little effort and without having to take any action. The proposed system specifically aims at hydroponic cultivation, in contrast to the systems covered previously. Using Arduino, Raspberry Pi 3, and TensorFlow, a prototype for the growth of a tomato plant was created [30]. Machine-to-machine connection and autonomous, intelligent control of the hydroponic system is made possible by the use of IoT. The system as it was designed is complex enough to give the necessary control action for the hydroponic environment based on the numerous acquired input parameters [30].

7. Small and medium-scale food production using hydroponic

Small firms have sprung up as a result of indoor production, boosting regional economies. The profitability of an indoor urban vertical farming (IUVF)-based company to that of a greenhouse was compared [41]. Due to the greenhouse's high maintenance costs, the results demonstrate that IUVF is more profitable on a small- to medium-sized scale than it is in the greenhouse. In order to set up a hydroponics farm, it is important to take into account the cost of various pieces of equipment, such as heating, ventilation, air conditioning (HVAC), fans, irrigation systems, control systems, railways, and lighting. The adoption of hydroponics as a farming technique is significantly hampered by the enormous initial cost of the system [42]. Thus, it is crucial to create new, better goods and services to aid the hydroponic system. Such technologies must be scaleable to meet growers' needs, not just for large-scale operations but also for medium- and small-scale ones, taking into account the limited amount of space that may be used for farming. Hydroponics is essential for the establishment and might make a substantial contribution to achieving sustainable development goal 11 (SDG 11). However, it requires the creation and uptake of the right technology. A framework that integrates heterogeneous devices on various computing layers to monitor and optimize the production process was designed to describe some technologies suitable for indoor farming [43]. In addition, they created the AgroRobot, an aeroponics-based robot for growing microgreens. Additionally, a number of project accessories, like culture trays, are created utilizing 3D printing technology. Hydroponics industry applications of Agriculture 4.0 are already well underway. A graphical user interface for indoor hydroponic production of leafy species, for instance, was created by the open-source Mycodo Environmental Regulation System [10].

8. Interlink between agriculture 4.0 and hydroponics

Hydroponics fits well into the framework of Agriculture 4.0 as large corporations increasingly use developments in indoor vertical farming, artificial intelligence, and plant biology to create a wide range of products [44, 45]. Hydroponics has secured a key position in the development of future food production systems, supported by cutting-edge, creative technology and a strong scientific foundation to ensure high output. Hydroponics may be very helpful in achieving SDG 11 which includes sustainability and resiliency of urban areas. The challenge is to adapt these technical developments to medium- and small-scale businesses that are prevalent in urban and peri-urban populations. Due to technological breakthroughs in electronics, which have led to the adoption of equipment, temperature and moisture sensors, aerial imaging, and GPS, modern farms. Precision irrigation, pest control, plant disease identification, and production management were all made possible by the integration of ideas such as artificial intelligence, the Internet of Things, and big data into autonomous food production systems. This revolution, dubbed "Agriculture 4.0," promises to seamlessly combine agricultural practices with cutting-edge technology, such as sensors, gadgets, machinery, and information technology. Until ideas like eco-agriculture [46], agro photovoltaics [47], and precision agriculture [48] are included in global farming practices and culture, Agriculture 4.0 has not yet been widely accepted. In order to achieve it, agriculture and technology will need to create a common ground for both farmers and technologists. The first step is to understand the best uses of technology and demand innovations that address the true needs of the food supply and value chains in order to meet producer expectations for highly improved products, services, and processes to support sustainable and effective food production in urban and peri-urban settlements. Aerial and satellite photography, robots, temperature and moisture sensors, GPS technology, and other sophisticated technologies are just a few of the ones that are being used more and more to enhance the entire food value chain and make businesses safer, more efficient, and more environmentally friendly [49].

9. Conclusion

Agriculture is a key component of the Indian economy. As a result of rising food demand, rising labour costs, unfavorable environmental conditions, and dwindling agricultural land, indoor farming techniques like hydroponics and aeroponics are becoming more and more popular. Hence, there is no mention of soil-based on the conventional manner of growing plants. That indicates that plants can develop if certain conditions are met. In this context, the concept of hydroponic farming is introduced in the farming system. In this method, mineral fertilizer solutions in water serve as a solvent, allowing plants to absorb nutrients from the soil more effectively. A variety of flowers, vegetables, and herbs can be grown with hydroponics. Due to the advancement of IoT, farmers may now automate hydroponic culture. Monitoring of water level, pH, temperature, velocity, and light intensity can be done with the help of IoT. For instance, the water tends to freeze in some areas throughout the winter, which could entirely obstruct the agricultural process. The water temperature sensors on the hydroponics farm can identify the temperature loss and alert the farmer as necessary. Expanding the research area, and study period, and investigating alternative methods of a more cost-effective hydroponic farming system, farmers allow to innovate in the farming system at a lower cost to produce high-quality crops.

Acknowledgements

We express our gratitude to the editor and reviewers for their valuable feedback on the manuscript.

Conflict of interest

The authors declare no conflict of interest.

Contribution of the authors

Writing of the manuscript: SD and BM; Correction of the manuscript: AS and SD.

Author details

Suman Dutta[1], Bishal Mukherjee[2] and Ashutosh Sawarkar[1*]

1 Division of Genetics and Plant Breeding, Ramakrishna Mission Vivekananda Educational and Research Institute, West Bengal, India

2 School of Agriculture and Allied Science, The Neotia University, West Bengal, India

*Address all correspondence to: annu.sawarkar@gmail.com

References

[1] Yegul U. Development of an embedded software and control kit to be used in soilless agriculture production systems. Sensors. 2023;**23**(7):3706

[2] Hossain SMM, Imsabai W, Thongket T. Growth and quality of hydroponically grown lettuce (Lactuca sativa L.) using used nutrient solution from coconut-coir dust and hydroton substrate. Advances in Environmental Biology. 2016;**10**(4):67-80

[3] World Food and Agriculture, Statistical Yearbook. 2023. Available from: http://www.fao.org/documents/card/en/c/ca6463en

[4] Jones Jr JB. Hydroponics: A Practical Guide for the Soilless Grower. 2nd ed. Boca Raton: CRC Press; 2004

[5] Sharma N, Acharya S, Kumar K, Singh N, Chaurasia OP. Hydroponics as an advanced technique for vegetable production: An overview. Journal of Soil and Water Conservation. 2018;**17**(4):364-371

[6] Verdouw C, Tekinerdogan B, Beulens A, Wolfert S. Digital twins in smart farming. Agricultural Systems. 2021;**189**:103046

[7] Rubio-Asensio JS, Parra M, Intrigliolo DS. Open field hydroponics in fruit crops: Developments and challenges. In: Fruit Crops. Amsterdam, Netherlands: Elsevier; 2020. pp. 419-430

[8] Wootton-Beard P. Growing Without Soil: An Overview of Hydroponics. Farming Connect. Penglais, United Kingdom. 2019

[9] Rufi-Salis M, Calvo MJ, Petit-Boix A, Villalba G, Gabarrell X. Exploring nutrient recovery from hydroponics in urban agriculture: An environmental assessment. Resources, Conservation and Recycling. 2020;**155**:104683

[10] Velazquez-Gonzalez RS, Garcia-Garcia AL, Ventura-Zapata E, Barceinas-Sanchez JDO, Sosa-Savedra JC. A review on hydroponics and the technologies associated for medium-and small-scale operations. Agriculture. 2022;**12**(5):646

[11] Gianquinto G, Munoz P, Pardossi A, Ramazzotti S, Savvas D. Soil fertility and plant nutrition. In: Good Agricultural Practices for Greenhouse Vegetable Crops. Rome, Italy: Food and Agriculture Organization; 2013. p. 205

[12] Sardare MD, Admane SV. A review on plant without soil-hydroponics. International Journal of Research in Engineering and Technology. 2013;**2**(3):299-304

[13] Lu N, Shimamura S. Protocols, issues and potential improvements of current cultivation systems. In: Smart Plant Factory: The Next Generation Indoor Vertical Farms. Singapore: Springer; 2018. pp. 31-49

[14] Jackson K, Meetei TT. Influence of soil pH on nutrient availability: A review. Journal of Emerging Technologies and Innovative Research. 2018;**5**:708-713

[15] Savvas D, Gruda N. Application of soilless culture technologies in the modern greenhouse industry—A review. European Journal of Horticultural Science. 2018;**83**(5):280-293

[16] Sonneveld C, Voogt W, Sonneveld C, Voogt W. Nutrient solutions for soilless cultures. In: Plant Nutrition of

Greenhouse Crops. Dordrecht: Springer; 2009. pp. 257-275

[17] Raviv M, Krasnovsky A, Medina S, Reuveni R. Assessment of various control strategies for recirculation of greenhouse effluents under semi-arid conditions. The Journal of Horticultural Science and Biotechnology. 1998;**73**(4):485-491

[18] Passam H. Hydroponic Production of Vegetables and Ornamentals. Athens: Embryo publications; 2002. pp. 15-23

[19] Saraswathi D, Manibharathy P, Gokulnath R, Sureshkumar E, Karthikeyan K. Automation of hydroponics green house farming using IoT. In: IEEE International Conference on System, Computation, Automation and Networking (ICSCA). Pondicherry, India: IEEE; 2018. pp. 1-4

[20] Tavakkoli E, Rengasamy P, McDonald GK. The response of barley to salinity stress differs between hydroponic and soil systems. Functional Plant Biology. 2010;**37**(7):621-633

[21] Phutthisathian A, Pantasen N, Maneerat N. Ontology-based nutrient solution control system for hydroponics. In: 2011 First International Conference on Instrumentation, Measurement, Computer, Communication and Control. Beijing, China: IEEE; 2011. pp. 258-261

[22] Sonneveld C, Voogt W, Sonneveld C, Voogt W. Nutrient management in substrate systems. In: Plant Nutrition of Greenhouse Crops. Dordrecht: Springer; 2009. pp. 277-312

[23] Ferentinos KP, Albright LD. Predictive neural network modeling of pH and electrical conductivity in deep–trough hydroponics. Transactions of the ASAE. 2002;**45**(6):2007

[24] Samarakoon UC, Weerasinghe PA, Weerakkody WAP. Effect of electrical conductivity [EC] of the nutrient solution on nutrient uptake, growth and yield of leaf lettuce (Lactuca sativa L.) in stationary culture. Tropical Agricultural Research. 2006;**18**:13

[25] Ropokis A, Ntatsi G, Kittas C, Katsoulas N, Savvas D. Effects of temperature and grafting on yield, nutrient uptake, and water use efficiency of a hydroponic sweet pepper crop. Agronomy. 2019;**9**(2):110

[26] Al-Rawahy MS, Al-Rawahy SA, Al-Mulla YA, Nadaf SK. Influence of nutrient solution temperature on its oxygen level and growth, yield and quality of hydroponic cucumber. Journal of Agricultural Science. 2019;**11**(3):75-92

[27] Qi Q, Tao F, Hu T, Anwer N, Liu A, Wei Y, et al. Enabling technologies and tools for digital twin. Journal of Manufacturing Systems. 2021;**58**:3-21

[28] Barricelli BR, Casiraghi E, Fogli D. A survey on digital twin: Definitions, characteristics, applications, and design implications. IEEE Access. 2019;**7**:167653-167671

[29] Rasheed A, San O, Kvamsdal T. Digital twin: Values, challenges and enablers from a modeling perspective. Ieee Access. 2020;**8**:21980-22012

[30] Mehra M, Saxena S, Sankaranarayanan S, Tom RJ, Veeramanikandan M. IoT based hydroponics system using deep neural networks. Computers and Electronics in Agriculture. 2018;**155**:473-486

[31] Tatas K, Al-Zoubi A, Christofides N, Zannettis C, Chrysostomou M, Panteli S, et al. Reliable IoT-based monitoring and control of hydroponic systems. Technologies. 2022;**10**(1):26

[32] Tzounis A, Katsoulas N, Bartzanas T, Kittas C. Internet of things in agriculture, recent advances and future challenges. Biosystems Engineering. 2017;**164**:31-48

[33] Tzounis A, Katsoulas N, Bartzanas T, Kittas C. Internet of things in agriculture, recent advances and future challenges. Biosystems Engineering. 2017;**164**:31-48

[34] Wang J, Yang Y, Wang T, Sherratt RS, Zhang J. Big data service architecture: A survey. Journal of Internet Technology. 2020;**21**(2):393-405

[35] Wang J, Chen W, Wang L, Sherratt RS, Alfarraj O, Tolba A. Data secure storage mechanism of sensor networks based on blockchain. Computers, Materials & Continua. 2020;**65**(3):2365-2384

[36] Zhang J, Yang K, Xiang L, Luo Y, Xiong B, Tang Q. A self-adaptive regression-based multivariate data compression scheme with error bound in wireless sensor networks. International Journal of Distributed Sensor Networks. 2013;**9**(3):913497

[37] Liao MS, Chen SF, Chou CY, Chen HY, Yeh SH, Chang YC, et al. On precisely relating the growth of Phalaenopsis leaves to greenhouse environmental factors by using an IoT-based monitoring system. Computers and Electronics in Agriculture. 2017;**136**:125-139

[38] Codeluppi G, Cilfone A, Davoli L, Ferrari G. LoRaFarM: A LoRaWAN-based smart farming modular IoT architecture. Sensors. 2020;**20**(7):2028

[39] Rajalakshmi P, Mahalakshmi SD. IOT based crop-field monitoring and irrigation automation. In: 2016 10th International Conference on Intelligent Systems and Control (ISCO). Coimbatore, India: IEEE; 2016. pp. 1-6

[40] Trilles S, Torres-Sospedra J, Belmonte O, Zarazaga-Soria FJ, Gonzalez-Perez A, Huerta J. Development of an open sensorized platform in a smart agriculture context: A vineyard support system for monitoring mildew disease. Sustainable Computing: Informatics and Systems. 2020;**28**:100309

[41] Avgoustaki DD, Xydis G. Indoor vertical farming in the urban nexus context: Business growth and resource savings. Sustainability. 2020;**12**(5):1965

[42] Caputo S. History, techniques and technologies of soil-less cultivation. In: Small Scale Soil-less Urban Agriculture in Europe. Cham: Springer; 2022. pp. 45-86

[43] Gnauer C, Pichler H, Schmittner C, Tauber M, Christl K, Knapitsch J, et al. A recommendation for suitable technologies for an indoor farming framework. e & i Elektrotechnik und Informationstechnik. 2020;**137**:370-374

[44] Hati AJ, Singh RR. Smart indoor farms: Leveraging technological advancements to power a sustainable agricultural revolution. AgriEngineering. 2021;**3**(4):728-767

[45] Dutta S, Singh AK, Mondal BP, Paul D, Patra K. Digital inclusion of the farming sector using drone technology. In: Human-Robot Interaction - Perspectives and Applications. London, United Kingdom: IntechOpen; 2023

[46] Keating BA, Carberry PS, Bindraban PS, Asseng S, Meinke H, Dixon J. Eco-efficient agriculture: Concepts, challenges, and opportunities. Crop Science. 2010;**50**:S-109

[47] Weselek A, Ehmann A, Zikeli S, Lewandowski I, Schindele S, Hogy P. Agrophotovoltaic systems: Applications, challenges, and opportunities. A review. Agronomy for Sustainable Development. 2019;**39**:1-20

[48] Pathak HS, Brown P, Best T. A systematic literature review of the factors affecting the precision agriculture adoption process. Precision Agriculture. 2019;**20**:1292-1316

[49] Araujo SO, Peres RS, Barata J, Lidon F, Ramalho JC. Characterising the agriculture 4.0 landscape—emerging trends, challenges and opportunities. Agronomy. 2021;**11**(4):667

Chapter 4

Advanced Micro Irrigation Techniques

Pavan Kumar Reddy Yerasi, V. Siva Jyothi, K. Madhusudhan Reddy, B. Sahadeva Reddy and C. Nagamani

Abstract

Problem statement: Indian agriculture's food basket has grown like a giant elephant with the advent of the green revolution, and later on, the micro irrigation task force came into existence in India. With the growing population, food requirement has to shoot up. Hence, productivity should be increased. With the advent of micro irrigation techniques, the area under irrigation increased enormously, and the footprint of the crop yields reached a pinnacle, ranging from horticultural to field crops, *etc*. Objectives: Micro irrigation techniques are trekking towards doubling the farm income by increasing productivity, *i.e*., doubling the farmer's income, and increasing water use efficiency in a bi-directional mode as resource enhancement and judicious use of resources on the other. Traditional micro irrigation methods cost more for farm installation, which is a big financial task that hinders adoption and leaves farmers looking for government policy support. There is a dire need to address this issue; alternate, best, and most effective systems must be facilitated for the farmer at an affordable price. In this context, rain port irrigation and laser irrigation *viz*., are innovative, highly efficient conveyance systems with higher water use efficiency and are available in an affordable price range for small and marginal farmers to bring more area under micro irrigation to achieve more harvest in a wide range of crops. Critical numerical findings: These systems have an edge-cutting advantage over the existing micro irrigation systems (sprinkler and drip) in terms of cost, ease of operation, water use efficiency, *etc*. Furthermore, they can be adopted for a wide range of crops, like both agriculture and horticulture, and other micro irrigation systems with single-man handling operations. This type of system has advantages over other micro irrigation systems in terms of increasing the harvestable basket, water use efficiency, and water conservation. Conclusion: The laser irrigation and rain port irrigations are alternate to drip and sprinkler irrigations, respectively. They could be used in a wide range of crops with precise water delivery system, easy to handle, one farmer can irrigate the entire field, unlike in sprinkler irrigation where carrying and installation require 2–3 man labourers. The droplet size and the pressure with which it exerts on ground are less compared to sprinkler irrigation leads creating lower bulk density and high porosity, causing better root growth and nodule formation in groundnut. The availability of nutrients is higher in laser spray irrigation followed by rain port mini sprinkler system and sprinkler irrigation.

Keywords: micro irrigation, laser spray, laser drip, rain port irrigation, water use efficiency

1. Introduction

Micro irrigation was an untapped potential during the early twentieth century in the Indian agriculture scenario. The productivity of rainfed crops is always dictated by the quantity and pattern of rainfall received during the crop season. A farmer's livelihood is invariably linked with rain, particularly in drought-prone arid districts of Andhra Pradesh [1]. In the present scenario, the irrigation of the crops has changed to feed the root, not the crop, and in the future, the definition may change in a new direction [2]. However, a National Task Force Committee, appointed by the Government of India in 2003, has recommended that 69 million ha of the area be suitable for micro irrigation in India. A target of 14 million ha has been suggested for the Eleventh Five-Year Plan [3]. Initially, irrigation was done through flooding from 1950 to 2000, then it moved to micro irrigations such as drip and sprinkler, which ruled over a period of two decades, *viz*., 2006 to 2020. Now, the concept of irrigation has become more precise with the advanced methods of irrigation such as rain port irrigation (micro jet sprinkler), laser irrigation, and so on [2]. Precision farming practices, along with plasticulture technologies such as micro irrigation techniques, have proven to be a driving force for the enhancement of farmers' income through increased productivity and optimum utilization of various inputs.

With the advent of the task force on micro irrigation and government incentives, the area under micro irrigation has increased enormously in India, especially in a few other states like Andhra Pradesh, Maharashtra, Tamil Nadu, etc. In districts like Ananthapuramu of Andhra Pradesh, the harvested water can be given as supplemental irrigation to the crops to enhance yields. However, in the recent past, many studies have revealed that drip and sprinkler irrigation deepened the water table depth in a few parts of Maharashtra and southern states due to the free power supply and automated irrigation switching systems. Hence, an alternate method of micro irrigation must be addressed to sustain micro irrigation in the long run. The best alternate method of micro irrigation is laser irrigation, to drip and sprinkler irrigation methods. Micro irrigation technologies (MI) are being expanded horizontally in vast stretches across the length and breadth of the country, covering 3.56 million ha of the area under micro irrigation in the sampled 13 states [4]. The micro irrigation techniques expanded vertically from orchards to ornamental crops too. These technologies are promoted primarily as: (1) a means to save water in irrigated agriculture; (2) a strategy to increase income and reduce poverty; and (3) to enhance the food and nutritional security of rural households. Despite the reported significant economic advantages and the concerted support of the government and non-governmental organizations (NGOs), the current area under micro irrigation is expanding on a large scale, but the water use efficiency has not improved much across the country. Still, there is a dire need to develop advanced micro irrigation systems with higher water use efficiency.

Different advanced micro irrigation systems are to be discussed here under with ease of operation and improved water use efficiency.

2. Journey of Indian irrigation concept

The Vedas first reported about irrigation in mythology, and the earliest mentions of irrigation were found in the Rig Veda. It has mentions of "kupa" and "avata" wells from where the water is drawn by "varatra" (rope strap) and "chakra" (wheel) pulling "kosa" (pails) of water. Later on, a few of the experts quoted in the 1980s said about nurturing the soil with nutrients and water to harvest more.

3. Sprinkler irrigation

Sprinkler irrigation is a method of applying irrigation water that is similar to rainfall. Water is distributed through a system of pipes, usually by pumping. It is then sprayed into the air and irrigated across the entire soil surface through spray heads so that it breaks up into small water drops that fall to the ground. Sprinklers provide an efficient coverage for small to large areas and are suitable for use on all types of properties. It is also adaptable to nearly all irrigable soils since sprinklers are available in a wide range of discharge capacities.

3.1 Practical layout and applicability of the sprinkler for groundnut crop as example

a. Any pipe with ISI marking having a length of 6 m or 20 feet is sufficient for a sprinkler pipe.

b. Each sprinkler's raised head discharges approximately 1800 L of water per hour (lph).

c. The throwing radius of each sprinkler raised head is 12 m (operates at 2.5 kg cm^{-2} pressure normally).

d. In sprinkler irrigation, the arrangement should be placed in a triangular pattern to achieve higher uniformity, as depicted in **Figure 1**.

e. Each sprinkler covers an area of 144 m^2. Ten sprinklers could cover an area of 1440 m^2 (144 m^2 × 10 sprinklers), approximately one third of an acre.

f. To achieve 95% uniformity in a 1440 m^2 area, 16 pipes and 8 raisers are required. The irrigation requirement for groundnut is 450–500 mm on varied crop coefficients, evapotranspiration (ET), and plant water demands depending on the conditions.

Figure 1.
Laser spray irrigation system at the ARS Research farm, Ananthapuramu.

g. The number of irrigations required for the groundnut is 8–9, approximately. (9 irrigations × 50 mm each through flooding = 450 mm), such as sowing before flowering (30–35 days after sowing (DAS)) and pegging to pod formation (around 35–90 DAS). Six irrigations are required based on evapotranscription (ET), rainfall, and soil conditions; finally, one irrigation is required 15 days before harvest.

h. With the same principle and nine irrigations through sprinkler irrigation of 40 mm each, by maintaining the acibenzolar-S-methyl (ASM) in the root, we can save around 100 mm of water for one hectare.

3.2 SWOT in Sprinkler irrigation

See **Table 1**.

Strengths	**Weakness**
Wide adoptability	Difficulty in conveyance to other fields
Less automatization	More leakages in conveyance
More flexibility	Operated at few stages only
	Less precise
	Not suitable to operate at all stages
Opportunities	**Threats**
More scope to adopt in large areas	Excess discharges may cause ground water depletion
Well accepted and adopted in farming community	Clogging is the major problem

Table 1.
Strengths, weakness, opportunities and threats of drip irrigation system.

4. Drip irrigation

Drip irrigation is a type of micro irrigation system that has the potential to save water and nutrients by allowing water to drip slowly to the roots of plants, either from above the soil surface or buried from below the surface. The goal of the said irrigation is to place water directly into the root zone and minimize evaporation. Two important things need to be known before the application of drip irrigation are how much water is to be applied and how much time is required to discharge the desired quantity of water.

4.1 SWOT in drip irrigation

See **Table 2**.

5. Laser drip irrigation

Laser drip irrigation, also known as precision or targeted irrigation, is an advanced technique of punching holes on the pipe by laser technology at definite intervals for

Strengths	Weakness
Suitable for a wide range of crops	More cost of installation
Automation can be adopted in wide range	Requires a sand filter
Premium quality of harvests can be reaped	Clogging is problematic
Easy to install, replace, and repair	Requires frequent maintenance
Opportunities	**Threats**
More scope to adopt in large areas	Excess discharges may cause groundwater depletion
High value crops yield more	More power consumption due to automation
Grow more with less	
Wider acceptance and adoptability	

Table 2.
Strengths, weakness, opportunities and threats of drip irrigation system.

the discharge of minute droplets to the plant roots. It involves the use of lasers to precisely deliver water directly to the plant roots, thus minimizing water wastage and ensuring efficient water use. Laser irrigation is an innovative alternative to drip irrigation.

Laser irrigation can significantly reduce water consumption by delivering water only where it is needed, unlike sprinkler irrigation, thus conserving water resources. The precise and uniform water distribution eliminates overwatering or underwatering issues. It requires less energy compared to sprinkler systems because it does not rely on pumping large volumes of water. Laser irrigation targets the plant roots, minimizing water availability to weed seeds and reducing weed growth.

5.1 Specifications of laser drip irrigation system

The available length for laser drip irrigation is 54 m, the available diameters are 16 and 20 mm, and the available thicknesses are 5, 8, and 0.4 mm. It can work at ultra-low pressure, *i.e.*, 0.1–0.2 kg cm^{-2}. The laser hole discharge is 4 liters per hour (lph), and the spacing between the laser holes is 40 cm. Row-to-row lateral drip spacing is 1.2 m (approximately 4 feet). The depth of application per hour is 8.33 mm. The time required for running 1 acre through laser drip is 36 min. In a laser drip system, there are no drippers; hence, there is less chance of clogging (**Table 3**).

5.2 Durability and cost of laser irrigation material

5.2.1 Suitability of the crops

It is suitable for wide-spaced crops to narrow-spaced crops, *i.e.*, groundnut, greengram, blackgram, maize, cotton, soybeans, vegetable crops, horticulture crops, etc.

Laser drip irrigation offers several advantages over traditional drip irrigation methods. Here are some key points:

1. Very fine droplets and precise water delivery to the root zone.

	Laser drip	Traditional drip irrigation
Operating pressure required at dripper discharge point (kg cm^2)	0.1–0.2	1
Dripper discharge (lph)	4	4
Dripper spacing in the lateral (cm)	40	40
Row-to-row spacing of lateral (m)	1.2	1.2
Depth of application (mm $hr.^{-1}$)	8.33	8.33
Clogging	Negligible	High
Maintenance	Low	High
Pump horsepower (Hp) required	Low	High
Electricity consumption	Low	High
Cost	Highly economical	Expensive

Table 3.
Comparison of laser drip versus traditional drip irrigation system.

2. Increased water use efficiency by significantly reducing water loss due to evaporation, runoff, and deep percolation.

3. Installation and removal are easy compared to drip irrigation.

4. Improves nutrient availability in the soil and increases nutrient uptake.

6. Laser spray irrigation

It is a revolutionary new irrigation system that is an alternative to sprinkler irrigation. Later pipes are punched with minute holes for the discharge of water in the form of sprays, which simulate light rainfall during the operation and run at low pressure. Micro-sprinkling hose irrigation has the unique advantage of reducing labour and input costs. The irrigation system uses low pressure to deliver water to micro-spray emitters through water pipes and tapes and adopts grouped multiple holes to emit water to soils, which markedly saves electricity costs in irrigation [5].

6.1 Specifications of laser spray system

The laser spray system consists of lateral laser pipes and subline from the main line of water discharge. The available lateral diameters of the laser line are 32 and 40 mm and it has a roll size of 100 m length. The wall thickness of the lateral line is 0.3 mm. The discharge of each lateral line is 172–175 lph per metre length. The wetting diameter of each laser punch is 12 m; however, the best results can be obtained at a 10 m distance with 100% overlapping. It can drizzle up to 5–6 feet (1.5–1.8 m) height depending on the operating pressure. The droplet size is minute to very minute and the pressure it exerts on the ground is less (0.5–1.0 kg cm^{-2}) compared to the

DOI: http://dx.doi.org/10.5772/intechopen.112509

	Laser spray irrigation	Sprinkler irrigation
Spacing between laterals	6 × 6 m	12 × 12 m
Discharge (lph)	170	1500
Pressure (kg cm^{-2})	0.7	1.5–2
Depth of application (mm h^{-1})	28–32	10
Cost per acre (INR acre^{-1}) approx.	20,000	25,000
Radius of operation	4–5 m	10–12 m

Table 4.
Comparison of laser spray versus traditional sprinkler irrigation system.

sprinkler irrigation droplet size. Hence, the bulk density and porosity of the soil are not changed during crop season (**Table 4**).

6.2 Durability and cost of laser spray material

Laser spray accessories are very cheap (approximately 16,000–20,000 INR acre^{-1}) compared to drip and sprinkler irrigation, can be affordable by every farmer, and have a life span of 3–5 years depending on the usage and maintenance of the farmer.

6.3 Suitability of the crops

Laser irrigation can be adopted for a wide range of crops, from leafy vegetables to onions, because it enhances the humidity and alters the microclimate for better yields in the summer. In the hilly terrains of the Western Ghats, *viz.*, Ooty and Munnar areas, temperatures reach a little higher and require cool climates and higher humidity. In that context, laser spray lowers air temperatures by altering the microclimate. Laser irrigation is suitable for a wide range of field crops, *viz.*, cereals, pulses, and oilseed crops. Sprinkler irrigation may affect flowering, pollination, fruit set, etc., in certain crops and may be replaced with laser irrigation. Even for horticultural crops and greenhouses, it can be useful. As the droplet size is finer and discharge is just like drizzle, it infiltrates to more depth compared to sprinkler irrigation. The bulk density of soils under this system of irrigation will be less when compared with sprinkler irrigation, facilitating for easy penetration of pegs into soil.

Micro-spraying irrigation has great advantages in realizing the high efficiency and water saving of pipe irrigation systems. In recent years, the micro-spraying irrigation system has been widely adopted in cereal crops such as wheat and maize (**Figures** 2–5) [5].

6.4 Feasibility for adoption

1. The cost is a bit cheaper compared with the other micro irrigation systems.

2. It can be easily laid out and can be wound up too in off seasons.

3. Mostly, it was manufactured with high-density polyethylene/low-density polyethylene (HDPE/LDPE), which is easily flexible in operation.

Figure 2.
Laser pipe with punched laser holes.

Figure 3.
Saddle fitting.

4. The rusting problem can be avoided.

5. Higher water uses efficient and better alternate conveyance system.

6. Damages/repairs can be done or rectified through puncture kit.

7. It can be operated at low pressures with high water use efficiency.

Figure 4.
Laser spray pipe fitting to subline with saddle fitting.

Figure 5.
Laser end cap.

8. As the droplet size is minute, it has a greater efficiency and more water savings.

9. In short time, more discharge can be grounded.

10. Clogging problem can be avoided and even the runoff surface collection can be used for the conveyance in this system.

11. Can be utilized in all stages of the crop including flower and pod development stages also as the water jet does not force much.

6.5 Shortfall in adoption

1. Being manufactured with HDPE/LDPE, it can be prone to damages.

2. The HDPE material is so thin that it can face the rodent's damage

3. Lack of government policy on this system for the rapid adoption.

6.6 SWOT in laser irrigation

See **Table 5**.

Strengths	**Weaknesses**
Operated at lower pressures	New innovation—Difficulty to adopt
Higher water use efficiency	
Lower cost than other micro irrigation systems	
Easy to lay and windup	
Opportunities	**Threats**
Feasible to operate at all stages of crop	LDPE is prone to damage easily
No damage to fruit and flowering	Requires more skills to operates
Operates at lower pressures with high discharge	
No issue of clogging	
Extra filters are not required	
Less maintenance	
Any type of water can be used	

Table 5.
Strengths, weakness, opportunities and threats of laser spray irrigation system.

7. Rain port Irrigation system

A rain port system is an advanced version of sprinkler irrigation with a discharge rate of 800 liters per hour and uniformity of 90–95%. It is made with polyvinyl chloride (PVC) or HDPE at a low cost. Rain port sprinkler systems are mini-irrigation systems, *i.e.*, laterals and sprinklers can be easily shifted from one place to another. Reinstallation of the system is also easy and consumes less time and labour. Depending on the pressure and recharge capacity of a bore well, the laterals are spaced at a distance of 6–10 m.

It overcomes all the limitations of conventional sprinkler irrigation systems and yet meets the high standards of effective irrigation principles such as

1. High distribution uniformity.

2. Controlled application rate.

3. Gentle precipitation, low droplet impact on soil structure, and no damage to foliage.

4. Short irrigation cycles to provide optimal growing conditions with highly accessible water and nutrients in a controlled wetted and aerated soil profile.

In a rain port irrigation system, flexible polyethylene tubes are used as laterals, and high-performance, low-weight plastic sprinklers are connected to these tubes using easily detachable connectors. Sprinklers are fixed on mild steel (MS) raiser rods or bamboo sticks can also be used to reduce the cost of raiser rods.

7.1 Specifications of rain port system

a. Rain port system is made with the linear low-density polyethylene materials and is easily flexible and suitable to transport and layout.

b. The operating pressure required for the rain port system is 1.5 kg cm^2.

c. Throwing radius of each rain port sprinkler is 10 m (operates at 1.5 kg cm^2)

d. The distance between each rain port sprinkler for the effective uniformity is 8 m × 8 m

e. Each rain port sprinkler covers around 64 m^2 area × 10 sprinklers = 640 m^2). For 1 acre with the rain port system with 8 m × 8 m spacing, it requires 63 rain port mini sprinklers.

f. The discharge of each rain port sprinkler is around 540 L h^{-1}.

g. Depth of irrigation achieved with one rain port sprinkler is 2.7 mm h^{-1}.

h. The available diameter sizes of the rain port sprinklers are 25 and 32 mm.

i. The minimum and maximum operating pressures required for the rain port irrigation system are 2–4 kg cm^2 (**Table 6**).

Compared to the sprinkler system (10 m), the throwing radius is a little bit lower, *i.e.*, 9 m in the rain port system. The discharge rate of each rain port came down to one third of the sprinkler discharge to increase the water use efficiency and also to operate even under low water availability. The operational pressure for the rain port system is 1.5 kg cm^2, which is 0.5 kg cm^2 lower than that of the normal rain port sprinkler system. Though the discharge is lower than the sprinkler irrigation system, the depth of application is 8 mm per hour compared to the sprinkler, *i.e.*, 10 mm per hour. In the rain port system, the distance between the laterals is 9 m, whereas in the sprinkler irrigation system, the spacing between the laterals is 10 m.

Rain port is more efficient with water saving up to 33% across the cropping season in different crops including dry seeded rice when compared with surface irrigation techniques.

Parameter	Rain port irrigation	Sprinkler irrigation
Spacing between laterals	9 × 9 m	129 × 12 m
Discharge (lph)	540	1500
Pressure (kg cm^{-2})	1.5	1.5–2
Depth of application (mm h^{-1})	8	10
Cost per acre (INR $acre^{-1}$) approx.	20,000	25,000
Radius of operation	10 m	10–12 m

Table 6.
Difference between the rain port and sprinkler irrigation system.

7.2 Suitability of crops

Rain port irrigation system can be suitable for the wide range of crops, *i.e.* groundnut, greengram, blackgram, bengalgram, and sesamum to green forage crops, etc., where sprinkler irrigation is suitable except at flowering.

7.3 Feasibility for adaption

1. The cost is a bit cheaper compared with the other micro irrigation systems.
2. These are made with LDPE/HDPE which eases the conveyance operation from one farm to others.
3. Higher uniformity can be achieved with lower operation pressure and low power consumption.
4. One person can install the system easily unlike sprinkler jet.
5. Can be operated at wide ranges of discharges with low power consumption.
6. Costly materials like brass/bronze can be avoided for the raiser rods/ discharge guns.

7.4 Shortfall in adoption

1. Being manufactured with HDPE/LDPE, it can be prone to damages.
2. The HDPE material is so thin that it can face the rodent's damage.
3. Lack of government policy on this system for the rapid adoption.

7.5 SWOT in rain port irrigation

See **Table 7**.
See (**Figures 6–10**).

Strengths	Weakness
Lighter in weight, handling, and installation	New innovation—Difficulty to adopt
More precise discharge and more water saving	
Easy to lay and wind up compared to sprinklers	
Opportunities	**Threats**
One man can install and operate	LDPE prone to damage easily
Operates at low pressures	Requires more skills to operate
More areas can be covered with available pressure	

Table 7.
Strengths, weakness, opportunities and threats of laser spray irrigation system.

Figure 6.
Saddle.

Figure 7.
Flash cap.

Figure 8.
High-density polyethylene (HDPE) 32 mm lateral with mini sprinkler raisers.

Figure 9.
Rain port irrigation system in groundnut.

8. Conclusion

Laser irrigation is in an infant stage in India. It can be studied in depth for a wide range of crops. The area under laser irrigation may be increased by the policy decision imparted to the existing micro irrigation methods, *viz.*, drip and sprinkler irrigation.

Figure 10.
Rain port irrigation system in sesamum.

Furthermore, it is high time to adopt better micro irrigation techniques, such as laser irrigation, for a wide range of crops to upgrade the farmer's economic status by increasing productivity. The rain port irrigation system has the advantages over the sprinkler system by way of ease of operation, field shifting, higher water use efficiency, lower discharge rates, and lower cost. It is the best system to be promoted in the future.

Author details

Pavan Kumar Reddy Yerasi[1]*, V. Siva Jyothi[2], K. Madhusudhan Reddy[3], B. Sahadeva Reddy[1] and C. Nagamani[4]

1 ANGRAU-Agricultural Research Station, Ananthapuramu, Andhra Pradesh, India

2 ANGRAU-Agricultural Research Station, Reddipalli, Andhra Pradesh, India

3 ANGRAU- Regional Agricultural Research Station, Tirupati, Andhra Pradesh, India

4 ANGRAU- S.V. Agricultural College, Tirupati, Andhra Pradesh, India

*Address all correspondence to: y.pavankumarreddy@angrau.ac.in

DOI: http://dx.doi.org/10.5772/intechopen.112509

References

[1] Reddy PK, Sahadeva Reddy B, Malleswara Reddy A, Radha Kumari C, Reddy BR. Irrigation management in Pigeonpea under rainfed Alfisols. Journal of Pharmacognosy and Phytochemistry. 2020;**9**(6):136-139

[2] Reddy PK, Sahadeva Reddy B, Siva Jyothi V, Ashok Kumar K, Malleswara Reddy A. Irrigation methods to crops-past, present and future. Indian Farmer. 2021;**8**(03):247-252

[3] Yella Reddy K, Satyanarayana TV. Micro-Irrigation Pays Rich Dividends-Experiences of Andhra Pradesh, India. Thailand: Royal Irrigation Department; 2016

[4] Chandrasekaran M and Suresh Kumar D. Micro Irrigation: Economics and Outreach in Tamil Nadu. 2012

[5] Wang S, Ji P, Qiu X, Yang H, Wang Y, Zhu H, et al. Effect of border width and Micro-sprinkling hose irrigation on soil moisture distribution and irrigation quality for wheat crops. Applied Sciences. 2022;**12**:10954

Section 2

Irrigation Applications

Chapter 5

Monitoring of Irrigated Lands of the Hissar Valley of the Republic of Tajikistan

Aliev Nozim Numonovich

Abstract

Most of the territory of the Republic of Tajikistan (93%) is occupied by mountain ranges and only 7% of the territory is lands suitable for agriculture. In this regard, the population density is very high and amounts to 60 people per 1 square kilometer of the area. The population at the beginning of 2019 was 5.3 million people, and in 2021, it was 9,537,645 people, an increase of about 80%. The demographic situation, the lack of suitable land for agriculture leads to the violation of the ecological balance that has developed in the last 30 years in Tajikistan, including on irrigated lands, which led to a decrease in crop yields by 1.5–2.0 times. One of the reasons for the decrease in productivity is the irrational use of land resources and anthropogenic impacts on reclamation lands. The intensification of the use of agricultural lands leads to their degradation, the development of desertification; a special place is occupied by irrigated lands, the problems of their conservation and improvement of their condition. The solution of these problems is based on objective and up-to-date information about the state and use of irrigated lands, as well as timely and reliable information received. In this regard, there is a need to identify and assess the existing environmental and socio-economic problems in the use of reclamation lands, the timely reconstruction of reclamation systems, the development of ways to solve the identified problems, as well as the development of methodological recommendations to justify the effectiveness of the implementation of state agricultural land management programs, taking into account regional characteristics. The existing information on the state and use of agricultural lands is contained in various organizations, is fragmented, of different scales and of different information. To use such information, it is necessary to transform it, which can lead to a significant loss of information, a decrease in its accuracy, and, consequently, its reliability. Modern geoformation technologies using remote sensing methods of the Earth at the present time can significantly improve the quality of research conducted when monitoring irrigated lands. In this regard, monitoring, analysis, assessment and forecasting of the ameliorative state of irrigated lands of the Republic of Tajikistan, and their conservation based on the development of criteria for environmental and socio-economic assessment of the use of reclaimed lands and the development of recommendations for the reconstruction of melioration systems based on the use modern geoinformation technologies and the use of aerospace images is relevant.

Keywords: monitoring, landmanagment, melioration system, irrigate, agricultural crop, GIS, Gissar Valley, digital map, information

1. Introduction

The main results of the work of the geoinformation system for managing irrigated lands of the Gissar region are reports on the use of reclaimed agricultural land, made according to standard forms, forecasts of changes in the reclamation and environmental and economic situation, and, most importantly, thematic maps (cartograms) in an easy-to-read graphical form that reflect information on the use and condition of irrigated lands. Farmers and dekhkan farms grow crops to meet the needs of the national economy in food. The main agricultural crops in the study area are: wheat, cotton, rice and vineyards.

The organization of monitoring of irrigated lands in the Gissar Valley using geographic information systems provides for the following steps:

Stage 1: Analysis of complex information on the state and use of natural environments on the territory of the Gissar Valley. The collection of information material on the object of study is carried out from various information sources, information is requested from the data bank of regional significance, multispectral space images are obtained from the Landsat 7 and Landsat 8 satellites, analysis of test data.

Stage 2: Mathematical analysis of multispectral satellite images on the territory of the Gissar valley. Analysis of multispectral satellite images from Landsat 7 and Landsat 8 satellites using ArcGis 12 software, determination of irrigated land areas in the Hissar Valley. Deciphering and vectorization of raster data for the study area.

Stage 3: Substantiation of the representativeness of indicators for assessing the state of irrigated lands. Establishment of negative processes arising in the Gissar Valley, grouping into the following groups: natural, technogenic, social. Analysis using software products "Statistica" v.10.0, JMP Statistic 15 and Minitab 19, on the basis of which an analysis of a set of indicators describing the state of natural environments in contact with irrigated lands is carried out: regression, correlation matrix. Determination of hidden factors affecting the state of irrigated lands and forecasting the development of changes in their state. Definition of areas for research.

Stage 4: Conducting field research. Inspection of the reclaimed area on the ground, search for interpretation parameters, visual analysis of the soil cover and crops of reclaimed lands, identification of areas with pronounced signs of salinization, control of intermediate interpretation, measurement of the depth of groundwater and sampling in a network of boreholes and wells, determination of the coordinates of reference observation points to correct data using modern technologies, GPS, UAVs, etc.

Stage 5: Organization of a single information space for monitoring irrigated lands in the Gissar Valley. It is envisaged to involve departmental services of the region, which carry out various types of monitoring, into a single information space.

Stage 6: Creation of digital thematic maps of the irrigated lands of the Gissar Valley, dekhkan farms in the form of: geographical (location, condition, connections of existing natural and social phenomena, cadastral information, etc.); environmental (pollution, infection, radioactivity, susceptibility to dangerous natural and natural-technogenic processes: mudflows, landslides, snow avalanches, landslides, earthquakes, etc.) and others; maps (including 3D terrain models); cartograms showing the required thematic characteristics of the factors of natural and anthropogenic impact.

Thematic maps are updated based on the collection of data obtained from various sources on various monitoring activities carried out on the territory of the Gissar Valley, formed on the basis of the proposed methodology for selecting the most representative monitoring indicators for irrigated lands in the Hissar Valley using software products "Statistica" v.10.0, JMP Statistic 15 and Minitab 19.

Thus, the introduction of monitoring of irrigated lands in the Gissar Valley and based on the use of GIS technologies for monitoring irrigated lands makes it possible to create maps directly in digital form according to the coordinates obtained as a result of measurements on the ground or during the processing of Earth remote sensing (ERS) and UAV data obtained from various sources. Implementation of the analysis of multispectral satellite images from Landsat 7 and Landsat 8 satellites using ArcGis 12 software into the process of monitoring irrigated lands. When creating digital maps in a GIS environment, emphasis is placed on creating a structure of spatial relationships between objects.

Digital maps serve as the basis for the production of conventional paper and computer maps on a paper basis and contain data and rules describing the position and spatial and logical relationships of objects in the territory of the Gissar Valley.

The aggravation of the ecological situation makes more and more relevant work on the consciousness of information bases, applied geographic information systems and the use of GIS technologies to solve a set of problems that arise in the field of nature management and environmental protection. In the field of monitoring irrigated lands using GIS, it is possible to solve the following main tasks:

- reflection of the current state of the land and the adjacent water area of the shelf zone by individual parameters or their possible combinations in the form of maps of various scales for various territorial units and levels;

- assessment of the state and dynamics of irrigated lands according to various parameters (erosion, salinization, soil pollution, acidity, etc.);

- assessment of the area and productivity of agricultural land;

- forecast of possible changes in the quality of irrigated lands;

- assessment of economic damage from pollution of irrigated lands by air, water and other sources;

- modeling of ecological processes on irrigated lands and on the territory of the Gissar valley;

- assessment of the condition of ameliorative systems and design of their reconstruction;

- forecasting and designing the use of irrigated lands, their protection.

Using the developed methodology, data were collected from ground control points to obtain orthophotomaps from multispectral satellite images obtained from the Landsat 7 and Landsat 8 satellites in order to compare spatial data with the available data. To implement this approach, all locations of soil samples were recorded using a Garmin 90 hand-held GPS navigator, which recorded more than 150 points covering all parameters for linking remote sensing data to this territory.

Classification and generalization of input data was carried out on the basis of geographic databases compiled by the author, taking into account the table of attributes and primary materials.

For further research, the author selected the irrigated lands of the dekhkan farm "Istiklol" of the Gissar district, located in Central Tajikistan. The total area of key sites is 150 hectares or 0.2% of the total monitoring area.

Field observations were carried out in determining the on-farm boundaries of land use; conditions of windbreaks; when taking soil samples to analyze its content in order to determine the degree of soil fertility in irrigated ecosystems of the study area.

Further processing was carried out on the basis of the developed model for creating monitoring maps for irrigated lands of the Gissar Valley using GIS and statistical data processing presented by "Statistica" v.10.0, JMP Statistic 15 and Minitab 19.

Issues related to anthropogenic impacts on the destruction of the ecological balance of irrigated potential on the process of degradation and water erosion of lands, which was obtained by us on the basis of field observations. Field surveys of dekhkan farms also provided data on the number of major crops produced on each farm.

Mapping. On the basis of the proposed methodology, we have compiled a number of thematic maps related to the objectives of the study, as well as obtained actual maps of the study area using GIS technologies.

Rasterization was performed when processing a satellite image in order to obtain a raster image, as well as preparing digital data for entering into a database.

The method allowed us to prepare more than 20 types of spatial data and obtain raster data from the resulting vector files. Data digitization was carried out using ARC GIS & Microsoft Excel software.

Vectorization. The process of interpretation of space images was carried out by us in order to determine the varieties of agricultural crops in the study area and a number of photogrammetric approaches to determine the parameters of the studied physical objects reflected in the images. All physical objects located in the study area were vectorized.

Vectorization was carried out on the basis of Arc Catalog in Arc GIS program using the object editing method. As a result, vector files were obtained for all objects of study.

Mathematical and statistical methods. Mathematical calculation methods make it possible to determine the number of studied phenomena in a particular area. We needed to use this method when calculating the number of row crops and other physical objects, their location, area, which are reflected in satellite images.

When using statistical methods, we can write that all tabular data and demonstrative-schematic data were obtained on the basis of statistical calculations of data recalculations. The most informative indicators were determined using statistical methods and approaches.

Laboratory methods. The objectives of the study required the analysis of soil content in key areas in order to find out the proportion of soil fertility, as well as soil composition to increase the productivity of irrigated lands. Using laboratory analyses, soil samples were analyzed from three farms in the study area.

Thus, it can be considered a key indicator of the possibility of using this technique - this is the achievement of the goal, which is the implementation of state monitoring of the state and use of irrigated lands of the Gissar Valley using maps created using geoinformation technologies based on Arc Catalog in Arc GIS.

Based on this, we can single out the main criterion - the creation of thematic maps, for which the main indicators are: vector data, raster data, text data, their verification and processing.

Comparison of the two methods for these indicators. The indicators for data processing according to the second scheme depend on the specifics of the state and use of land in different areas of the valley. Comparing two methods of creating maps for the implementation of state monitoring of the state and use of land, it can be noted that when using any of them, you can get the desired result - a thematic map that displays the state and use of land of any territory of interest.

But at the same time, it should be noted that when using statistical data processing in the second method, it is possible to achieve a better and more correct display of the real situation on the ground due to a double check for the reliability of the data - statistical and expert.

Also, when using statistical processing, it is possible to find dependencies between indicators, determine trends in the development of indicators, predict the development of observed indicators, and make a management decision in advance to prevent the deterioration of the situation.

Conclusions. Thus, the authors in this article reviewed the analysis of the state of irrigated lands in the Gissar Valley based on the use of multispectral satellite images from the Landsat 7 and Landst 8 satellites using the ArcGis software. In the process of their analysis, maps were obtained for the NDVI index from 2010 to 2019, on the basis of which changes in the area of vegetation cover for the period 2010–2019 were obtained. And then, MNDWI maps were obtained from 2010 to 2019 of water coverage, and plots related to irrigated land were determined using the obtained vegetation and water coverage data.

On the basis of the received set, an analysis of the qualitative state of the lands in the Gissar Valley was carried out and it was found that in the territory of the valley, the development of processes of degradation of soil and vegetation covers is observed almost everywhere, which affects the efficiency of agriculture and causes the expansion of territories, the ecological state of which is problematic or even critical. Thus, one of the key directions in organizing the rational use of land in the region, as experience shows, should be land management of agricultural enterprises on an agrolandscape basis in combination with other works, which in turn requires up-to-date information about their condition and use, obtained as a result of monitoring of irrigated lands using modern information and telecommunication technologies and geographic information systems.

2. Development of theoretical provisions for monitoring irrigated lands in the Gissar Valley of the Republic of Tajikistan using GIS technologies

As noted earlier, land monitoring (Reimers, 1990) is based on a system of observations of various phenomena, conditions and processes, but a geographic information system is also very important for processing and predicting the state of the observed environment, given that in modern monitoring conditions, they receive a huge amount of various data that is almost impossible to process without the use of modern geoinformation technologies. Bearing in mind that information support for assessing the ecological state of irrigated lands should be based on obtaining up-to-date data on the state of the main components of the natural environment, the dynamics of changes and factors that affect the state of these lands.

Fomin et al. (1995), Derzhavin et al. (1999), Gostishchev et al. (2001); Biryulin and Manukyan (2002) note that the most important part of monitoring the state of irrigated lands is its information system.

Geographic Information System (GIS) - a software and hardware system capable of storing and using (showing, analyzing, managing) data describing objects in space, managed by special personnel, and the word "geographical" in this case means not so much "spatiality" or "territoriality", but rather the complexity and consistency of the research approach. The use of geographic information systems allows you to quickly receive information on request and display it on a map basis, assess the state of the ecosystem and predict its development.

GIS uses a special type of information - spatial (geographic) and associated databases, this information can be social, political, environmental or demographic, that is, any information that can be displayed on a map.

GIS is the best way to store information about land or sea. GIS can help make management more efficient, promote scientific work and the protection of natural resources, which is carried out for all territories, regardless of their area.

In recent years, geoinformation systems and WEB technologies have been actively used for processing aerospace images, accumulating and transmitting data.

The introduction of GIS systems solves a number of problems: maintaining a centralized accounting of agricultural lands, their inventory, monitoring the state and use, supervision of their use, preparation of analytical information; entering information into the information system for agricultural land management; use of Earth remote sensing data for monitoring the state of crops; accounting of land plots from agricultural land and land of other categories; preparation of information with aggregation at various levels, including the level of the Republic; providing access to information provided on the basis of state information resources on agricultural lands based on the use of GIS technologies. In fact, this system demonstrates an integrated approach to monitoring agricultural land. An important advantage of this approach is the possibility of creating a single Internet resource and providing access to geospatial data to interested parties [1].

It is important to note that the collection of information on agricultural lands in all the described systems is carried out mainly in the context of the state, regional, district levels, except for individual agricultural enterprises, and even more so crop rotation fields.

Geoinformation technologies provide ample opportunities in the development of irrigated land management systems. GIS allows structuring information on the geographical (spatial) position of objects - data sources, which is of particular importance for managing the state of not only irrigated lands, but also the environment.

Nowadays, one of the urgent tasks of agricultural science is to improve the quality and scientific validity of irrigated agriculture management, including through the development and use of modern software and information tools to support decision-making. Their main task is to provide all persons making decisions on the management of the reclamation complex with detailed, complete, timely and high-quality information in a form convenient for human perception, to ensure its mathematical and heuristic processing. At the same time, it is very important to link information flows that are heterogeneous in origin, but spatially interconnected.

The scope of GIS technologies is land management, land cadasters; design, engineering surveys and planning in urban planning; thematic mapping; inventory and accounting of objects; terrain analysis; land transport navigation; air traffic control; geology; environmental monitoring; environmental management; natural resource management.

The use of GIS technologies for monitoring irrigated lands makes it possible to create maps directly in digital form according to the coordinates obtained as a result of measurements on the ground or during the processing of remote sensing materials. When creating digital maps in the GIS environment, the emphasis is on creating a structure of spatial relationships between objects, the concepts of exact and inaccurate coincidence of boundaries are clearly distinguished, it is easy to use previously digitized boundaries when creating adjacent objects, including when working in other industries, it is easy and relations of connectivity, neighborhood, adjacency, nesting, intersection, and other spatial objects necessary for solving a wide range of analytical and practical problems are fixed explicitly).

Digital maps serve as the basis for the production of conventional paper and computer maps on a solid substrate and contain data and rules that describe the position and spatial and logical relationships of terrain objects [2].

Since the adaptation of all components of irrigated farming systems requires complete and systematic information about the properties of soils, groundwater, irrigation network, etc. for specific fields of the economy, the information support should be based on local integrated GIS monitoring [3].

A natural and necessarily arising problem of monitoring agrolandscapes is the problem of transition from point-based data on the state of indicators of the ameliorative state of lands and their soil fertility to digital models of the spatial distribution of these indicators in the layers of digital maps. At present, many tools have been developed for constructing spatial distribution models using interpolation and approximation of point data. Our task was to select the best methods and develop algorithms for their application in the specific conditions of irrigated lands in Tajikistan on the territory of the Gissar Valley [4].

Thus, the monitoring of irrigated lands is created as a multi-level observation system, including complexes of ground and remote (including aerospace, from unmanned aerial vehicles) observation methods:

- Detailed - examination of the state of land on the objects included in the irrigation system, the system of water protection facilities.
- Local - ground and aerial photography, including from UAVs (local observations and forecasts).
- Regional - aerospace survey (regional observation forecasts).

The monitoring program for irrigated lands must necessarily include primary information on the state and use of lands, the main criteria for this level are the importance and completeness of the information collected, such information can be obtained at the local level [5]. **Figure 1** shows the hierarchical structure of irrigated land monitoring using modern technologies.

Monitoring of the state of irrigated lands is carried out during complex work in the natural and technogenic systems of these lands, the main task of which is to obtain reliable and sufficient information about the factors that have a negative impact, the nature of their flow on irrigated lands and adjacent natural environments located in different landscape conditions under the actual use of land for agricultural production.

The monitoring program for irrigated lands should take into account the specifics of both the natural, economic, and socio-economic conditions of the Gissar Valley region, which are not repeated in other regions. Irrigated land monitoring information should be highly informative, meet the requirements of objectivity, reliability, accuracy, comparability, efficiency and availability [6, 7].

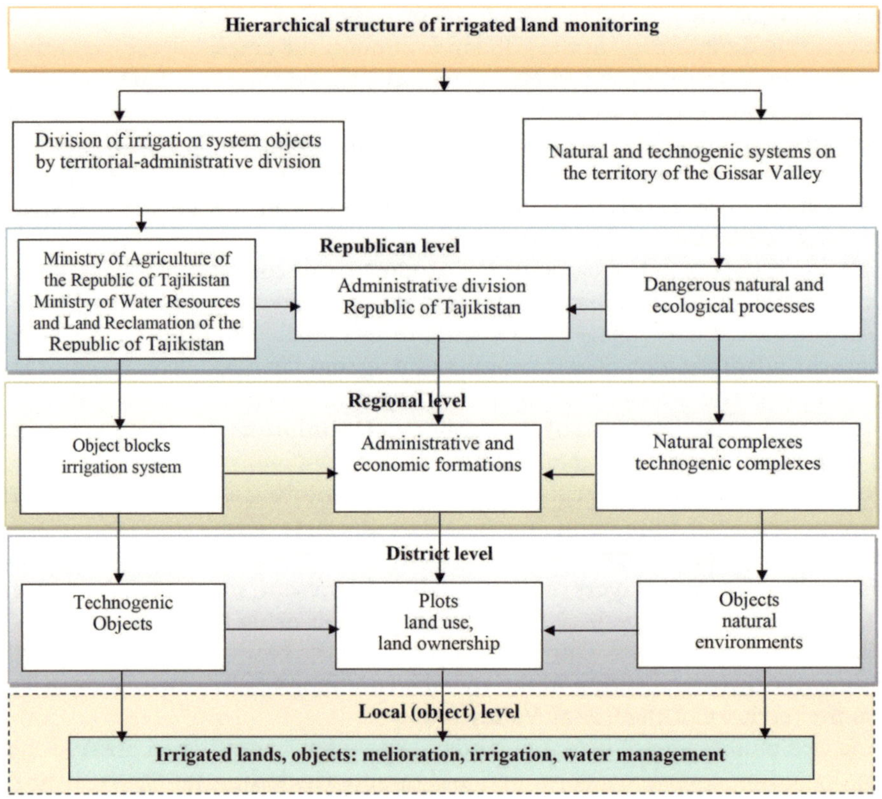

Figure 1.
Hierarchical structure of monitoring of irrigated lands in the Gissar Valley in the Republican system (compiled by the author).

The monitoring system of irrigated lands allows planning and predicting the use of lands and adjacent environments, taking into account their further safety, taking into account those degradation processes in irrigated agricultural landscapes that lead to a deterioration in their ameliorative state, are groundwater rise and secondary salinization.

The information base for monitoring irrigated lands is various thematic maps that serve to study the current state, identify disturbances and predict the state of lands and adjacent environments using regional features and informative indicators for monitoring irrigated lands in the natural-technogenic system of the Gissar Valley.

An important component for monitoring irrigated lands in the natural-technogenic system is the determination of the boundaries of irrigated lands, established as a result of work on the land cadaster, taking into account protected and sanitary protection zones (**Figure 2**).

Spatial modeling of the use and condition of irrigated lands is carried out using a geographic information system, which includes the necessary set of basic objects of water management and melioration on irrigated lands and thematic layers, which include: a complex of engineering structures directly related to melioration and irrigation, for example, such as: ameliorative facilities; irrigation systems, mud storages; land reclamation sites; groundwater intakes; enterprises of motor transport and special vehicles; land resources, as well as other components of the environment (atmospheric air, soils, surface and ground waters, vegetation).

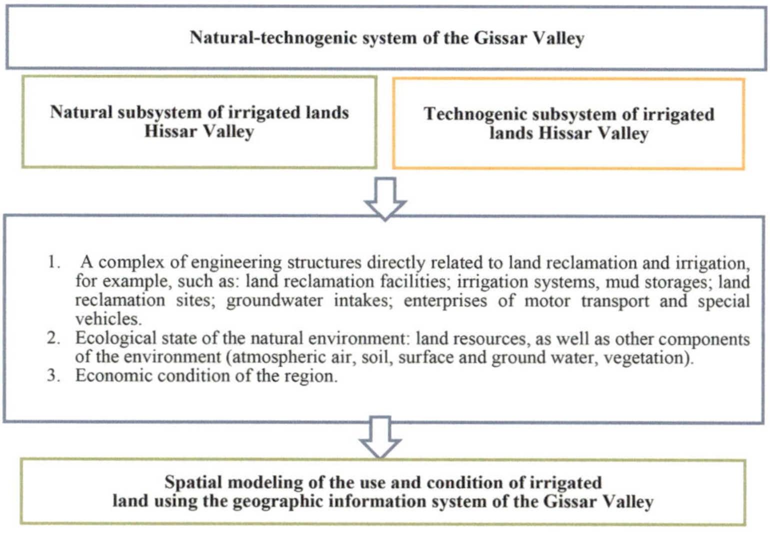

Figure 2.
Geoinformation system in the structure of modeling the natural-technogenic system of the Gissar Valley (compiled by the author).

From this it follows that one of the most important components of a digital map are layers that reflect the level of groundwater and salinity of irrigated lands and adjacent areas. In this connection, requirements are imposed on the information system, which should provide the following indicators: (1) correspond to the organizational structures of the exploitation of irrigated lands; (2) reflect the structure, properties and relationships of the observed objects and processes occurring in them; (3) represent reclamation lands as an integral part of such an element of the natural environment as an agrolandscape; (4) ensure the security and confidentiality of information, as well as free access to information users; (5) have organizational, software, technical, mathematical, methodological, linguistic, metrological and legal support; and (6) provide promptly and in an accessible form information (digital or paper forms) for the entire observation period for various analysis and forecasting models.

In many cases, the land assessment procedure involves the use of not only soil, but also other indicators characterizing the relief (steepness, slope exposure), vegetation, hydrothermal regime, parent rocks, hydrographic network, as well as socio-economic conditions. Therefore, an objective assessment of land, taking into account all factors, is carried out on the basis of GIS technologies and its result can be presented in the form of thematic maps.

The software companies ESRI (Environmental Research Institute, USA) and ERDAS provide an excellent basis for the implementation of a full-featured land use management system of the subject (monitoring at the state level). This system involves deep analytical processing of large volumes of data of different scales, as well as interfacing with complex software applications.

From the above examples, it can be seen that the efficiency, integration of environmental, socio-economic, biomedical, management information on a single geographical basis in conjunction with mathematical analysis tools distinguish geographic information systems from a number of existing applications. This versatility explains the attractiveness of their use for specialists in almost any industry, including the field of land reclamation.

If we talk in general about the use of information technologies in agroecological research, then it must be said that the creation of databases (Database) and GIS has been the basis of theoretical and applied research in the geosciences for many years. Their main purpose is to provide a spatial analysis of the placement, relationships, dynamics and other relationships of spatial objects. In combination with modern mathematical methods for processing thematic data, databases open up new ways to quantitatively study the patterns of functioning of bio-geosystems at different spatial levels. The introduction of new geoinformation technologies into the theory and practice of research is especially important when creating and monitoring irrigated agrolandscapes for the Republic of Tajikistan.

Based on the foregoing, let us consider the general theoretical provisions of monitoring irrigated lands. Since, as already discussed, the ameliorative state of lands and the development of new irrigated lands, together with natural objects, are a complex natural and technogenic system, the monitoring of these lands must be considered in conjunction with other types of monitoring: the state of water bodies and groundwater; atmosphere; soils; vegetation; surface waters; functioning of reservoirs of irrigation importance; relief, etc.

Information support is based on obtaining up-to-date data on the state of the main components of the natural environment, the dynamics of changes and factors that affect the state of irrigated lands.

The structure and content of the monitoring of irrigated lands at each specific facility will be largely determined by the complexity of the facility itself and the complexity of its involvement in its interaction with natural resources. In accordance with the tasks set, the Program for local monitoring of irrigated lands includes the following sections and areas of work [8]: (1) Determining the purpose and objectives of monitoring irrigated lands. (2) Compilation of the initial passport of the territories occupied by land reclamation facilities. (3) Determination of indicators and regulations for observations that take into account the cumulative impact of anthropogenic impact on land resources and adjacent environments. (4) Development of an observation network and means of observation. (5) Obtaining and processing monitoring information and (6) Assessment of the state of lands and adjacent environments.

In the course of research, the author developed a model for monitoring irrigated lands in the Gissar Valley (**Figure 3**).

The presented model includes the following stages: study of the methodological base for land monitoring by classification criteria; identification of natural and geographical features of the Gissar valley; determination of factors of natural and technogenic processes that have a negative impact on the ecological state of irrigated lands based on the use of the software product "Statistica" v.10.0; determination of the most informative indicators for monitoring irrigated lands in the Gissar Valley; development of a methodology for the formation and updating of information on the state and use of irrigated lands in the Hissar Valley, obtained from open sources, within the established boundaries of irrigated lands, taking into account the land use zone, engineering reclamation and irrigation object, water management objecties, as well as the natural environment.

Based on the characteristics of the object under study, the scheme of the irrigated land monitoring model algorithm proposed by us includes the following levels: regional, local and detailed. For irrigated lands, a natural-technogenic system is determined, taking into account the established boundaries in accordance with the land cadastre and the zone of technogenic influence on adjacent natural complexes, observation parameters, methods and means of observation, as well as the frequency

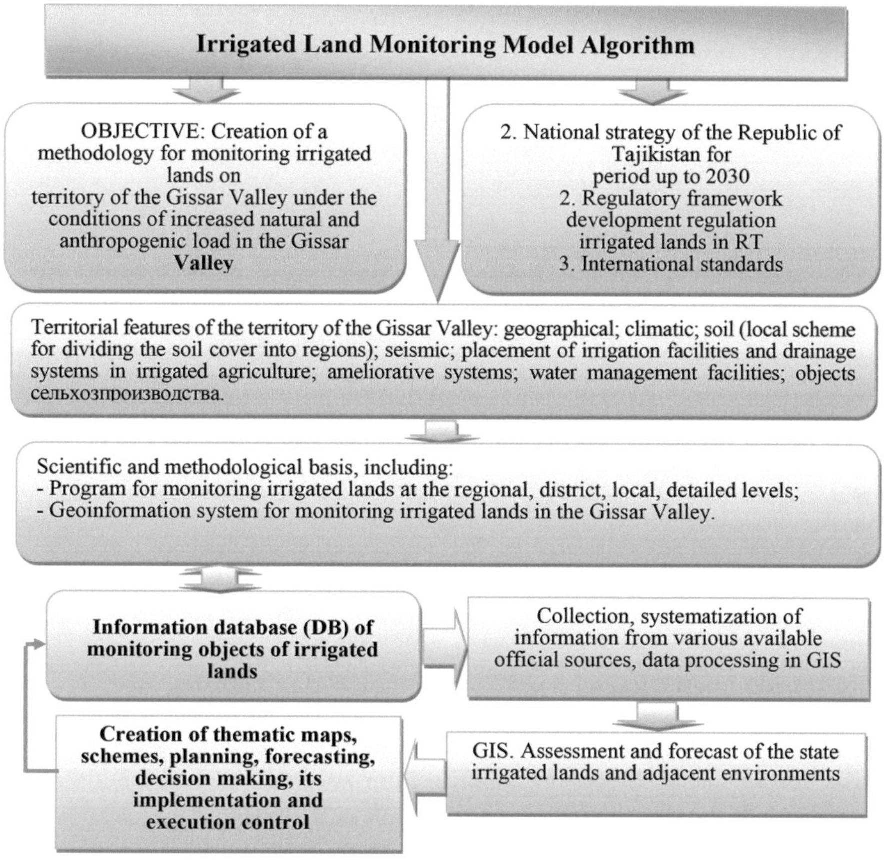

Figure 3.
Scheme of the algorithm of the irrigated land monitoring model using geographic information systems (compiled by the author).

of observation, the observed network, the algorithm for processing and issuing the requested information.

The monitoring program for irrigated lands is developed in such a way that it is possible to ensure hierarchical subordination according to the main levels of monitoring: regional, subregional, local and detailed, taking into account natural regional, local features and technical features of reclamation systems (**Figure 4**).

Thus, the main principles of building a monitoring system for irrigated lands should be the following:

- complexity, includes ground and remote means of monitoring the state of lands of reclamation systems and adjacent territories;

- openness of information necessary for monitoring both the state of irrigated lands and natural environments, and the state of engineering structures, making management decisions, monitoring the implementation of decisions;

- multi-level nature, taking into account the structure of objects and ensuring the sustainable development of agricultural production areas;

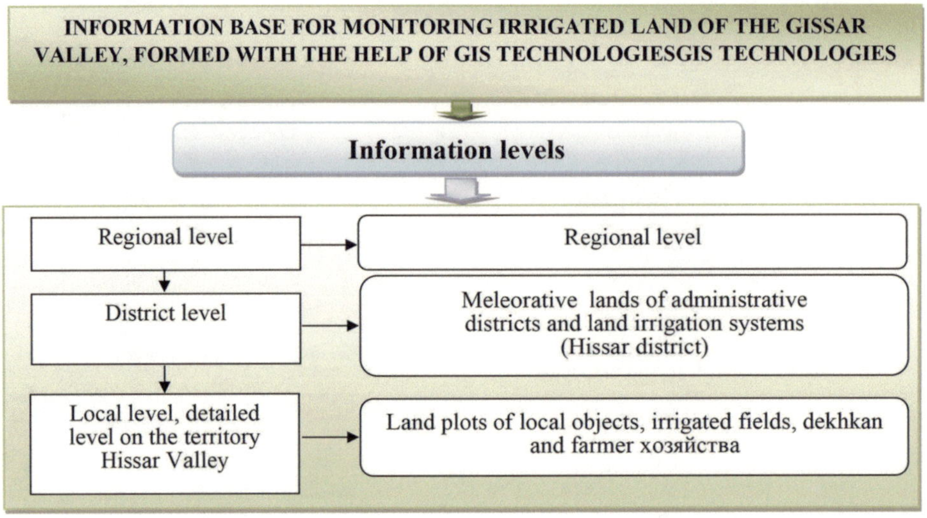

Figure 4.
Scheme of hierarchical subordination of the main levels of monitoring of irrigated lands using GIS technologies (compiled by the author).

- information interaction with other environmental monitoring systems at all levels;
- strict observance of the requirements of international standards of environmental protection in the sustainable development of agriculture;
- obligatory formation of complex information (cadastral, ecological state of irrigated lands and the environment, technical state of reclamation, irrigation and other water management systems).

Information sources in this case are all available official statistical data and data from the authorities of municipal and regional entities. The most detailed is monitoring at the level of an irrigated land plot, the territory of which is established within the boundaries, taking into account the buffer zone and adjacent territories.

Based on the collected information material using various sources, it is processed using geographic information systems, which allow for an objective assessment of the ecological and reclamation state of irrigated lands at the local level, identify ongoing negative processes and give objective recommendations for their prevention or elimination of consequences [9].

The cumulative assessment of the ecological and reclamation state of irrigated lands consists of the following stages:

Stage 1 - preparatory includes:

- selection and acquisition of satellite images;
- collection of information on the problem under study;
- a selection of computer components and methods for performing identification;

- familiarization with the explicit identification attributes and indicators of the object;
- intermediate recognition and analysis of features;
- definition of areas for research.

Stage 2 - field studies of selected areas using Landsat multispectral images, the selection of images is carried out depending on the task being solved, which corresponds to the maximum determination of the state of the study area for the annual period.

Stage 3rd - office processing using GIS.

Information collected during field research and interpretation of images is modified in GIS using the ArcGIS online program. Thus, the GIS "Gissar irrigation system" will contain the following data:

- about the object: areas, boundaries, terrain, hydraulic structures, etc.;
- according to available soils and growing vegetation: type, soil properties, level and mineralization of groundwater, names of species of plant organisms, etc.;
- obtained as a result of work on the ground: areas in the fields with pronounced processes of soil salinization; places where soil samples were taken, etc.;
- obtained after the processing of satellite information: areas of swamping and salinization of lands.

Thanks to the use of geographic information systems, a combination of heterogeneous information can be achieved to conduct a qualitative analysis of the ecological and reclamation state of irrigated lands.

3. Development of a model for creating maps of the state of irrigated lands in the Gissar Valley based on the developed methodology for using GIS technologies and software products

Using the developed methodology, data were collected from ground control points to obtain orthophotomaps from multispectral satellite images obtained from the Landsat 7 and Landsat 8 satellites in order to compare spatial data with the available data. To implement this approach, all locations of soil samples were recorded using a Garmin 90 hand-held GPS navigator, which recorded more than 150 points covering all parameters for linking remote sensing data to this territory.

Classification and generalization of input data was carried out on the basis of geographic databases compiled by the author, taking into account the table of attributes and primary materials.

For further research, the author selected the irrigated lands of the dekhkan farm "Istiklol" of the Gissar district, located in Central Tajikistan. The total area of key sites is 150 hectares or 0.2% of the total monitoring area (**Figure 5**).

Field observations were carried out in determining the on-farm boundaries of land use; conditions of windbreaks; when taking soil samples to analyze its content in order to determine the degree of soil fertility in irrigated ecosystems of the study area.

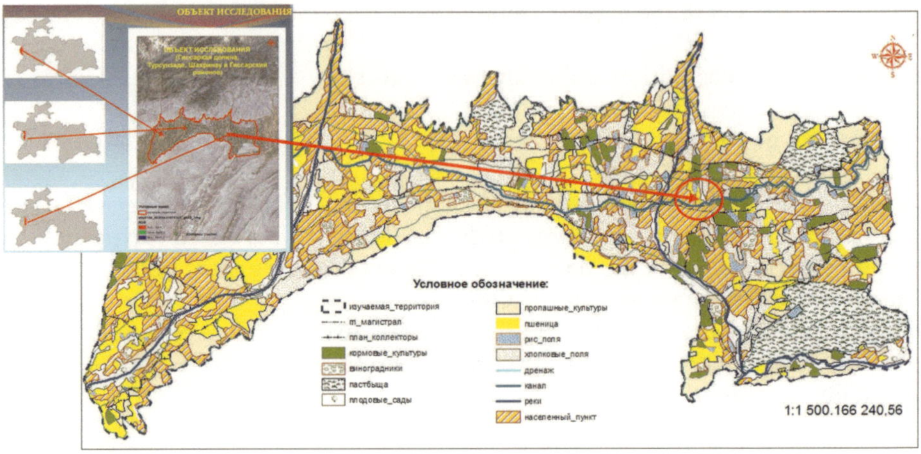

Figure 5.
Map of general land use in the study area (Tursunzade, Shakhrinau, Gissar districts) (compiled by the author).

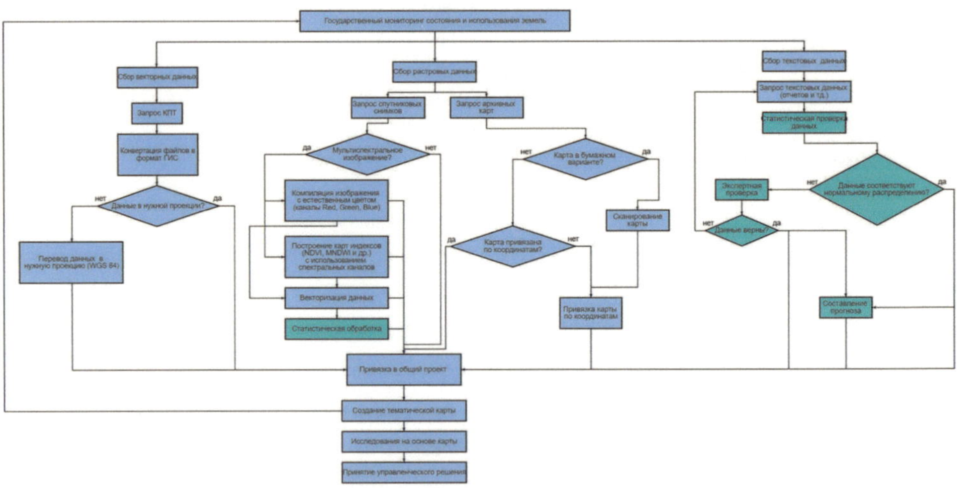

Figure 6.
Scheme for creating monitoring maps of irrigated lands in the Gissar Valley using Minitab 18.

Further processing was carried out on the basis of the developed model for creating monitoring maps for irrigated lands of the Gissar Valley using GIS and statistical data processing presented ("Statistica" v.10.0, JMP Statistic 15 and Minitab 19) (**Figure 6**).

On the basis of the studies performed, we proposed the following organization of work to study the state of irrigated lands on the territory of the Gissar Valley using geoinformation technologies, consisting of the following stages:

1. **Field** work was carried out to determine the on-farm boundaries of land use, the state of windbreaks, to take soil samples and analyze its content in order to determine the degree of soil fertility in the irrigated ecosystems of the study area.

   Issues related to anthropogenic impacts on the destruction of the ecological balance of irrigated potential on the process of degradation and water erosion of lands, which was obtained by us on the basis of field observations. Field surveys of dekhkan farms also provided data on the number of major crops produced on each farm.

2. **Mapping**. On the basis of the proposed methodology, we have compiled a number of thematic maps related to the objectives of the study, as well as obtained actual maps of the study area using GIS technologies.

3. Rasterization was performed when processing a satellite image in order to obtain a raster image, as well as preparing digital data for entering into a database. The method allowed us to prepare more than 20 types of spatial data and obtain raster data from the resulting vector files. Data digitization was carried out using ARC GIS & Microsoft Excel software.

4. Vectorization. The process of interpretation of space images was carried out by us in order to determine the varieties of agricultural crops in the study area and a number of photogrammetric approaches to determine the parameters of the studied physical objects reflected in the images. All physical objects located in the study area were vectorized.

   Vectorization was carried out on the basis of Arc Catalog in Arc GIS program using the object editing method. As a result, vector files were obtained for all objects of study.

5. Application of mathematical and statistical methods. Mathematical calculation methods make it possible to determine the number of studied phenomena in a particular area. We needed to use this method when calculating the number of row crops and other physical objects, their location, area, which are reflected in satellite images.

   When using statistical methods, we can write that all tabular data and demonstrative-schematic data were obtained on the basis of statistical calculations of data recalculations. The most informative indicators were determined using statistical methods and approaches.

6. Laboratory methods. The objectives of the study required the analysis of soil content in key areas in order to find out the proportion of soil fertility, as well as soil composition to increase the productivity of irrigated lands. Using laboratory analyses, soil samples were analyzed from three farms in the study area.

Thus, it can be considered a key indicator of the possibility of using this technique - this is the achievement of the goal, which is the implementation of state monitoring of the state and use of irrigated lands of the Gissar Valley using maps created using geoinformation technologies based on Arc Catalog in Arc GIS.

Based on this, we can single out the main criterion - the creation of thematic maps, for which the main indicators are: vector data, raster data, text data, their verification and processing.

Table 1 clearly shows the comparison of the two methods for these indicators.

The indicators for processing data according to the second scheme (**Table 1**) depend on the specifics of the state and use of land in different areas of the valley. Comparing two methods of creating maps for the implementation of state monitoring of the state and use of land, it can be noted that when using any of them, you can get the desired result - a thematic map that displays the state and use of land of any territory of interest.

But at the same time, it should be noted that when using statistical data processing in the second method, it is possible to achieve a better and more correct display of the real situation on the ground due to a double check for the reliability of the data - statistical and expert. Also, when using statistical processing, it is possible to find dependencies between indicators, determine trends in the development of indicators, predict the development of observed indicators, and make a management decision in advance to prevent the deterioration of the situation.

Data processing indicators Figure 1	(without statistical processing)	Scheme 2 (with statistical processing)
Collection of vector data		
Support for various formats	+	+
Support for various sources	+	+
Collecting raster data		
Support for various formats	+	+
Support for various sources	+	+
Collection of text data		
Support for various formats	+	+
Support for various sources	+	+
Vector data processing		
Correction of existing vector data	+	+
Drawing new vector data	+	+
Snap to project coordinate system	+	+
Raster data processing		
Image correction	+	+
Digitizing raster data	+	+
Snap to project coordinate system	+	+
Text data processing		
Data validation	—	+
Correlation analysis	—	+
Forecast for the development of indicators	—	+
Entering text data into the project database	+	+
Create a thematic map		
Output of the finished product	+	+

Table 1.
Comparison of methods for creating maps for monitoring the condition and use of irrigated lands in the Gissar Valley.

TIN is a form of vector digital geographic data that is built by triangulating a set of vertices (points). The vertices are connected by a series of edges and form a network 03 of triangles. Therefore, in order to model the TIN surface for the study area - alluvial marches, we first of all use the Global Mapper GIS to download data on the surface of the study area from the Advanced Spaceborne Thermal Emission and Reflection Radiometer to find the height points of the study area (**Figure 7**). Territory (**Figure 8**). Next, surface points with height data were displayed (**Figure 9**).

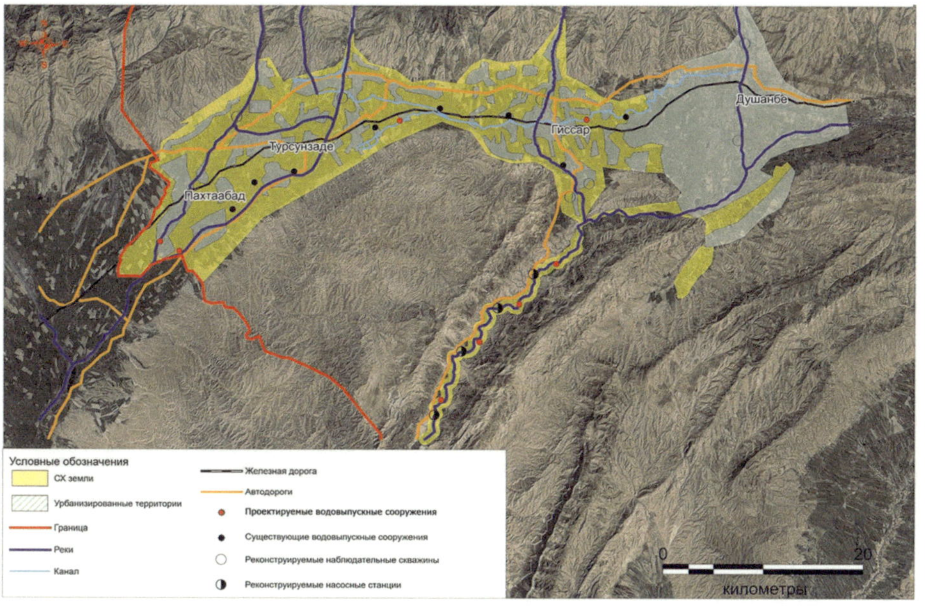

Figure 7.
Map of rehabilitation measures for the irrigation system of the Gissar Valley on the Landsat 8 satellite image.

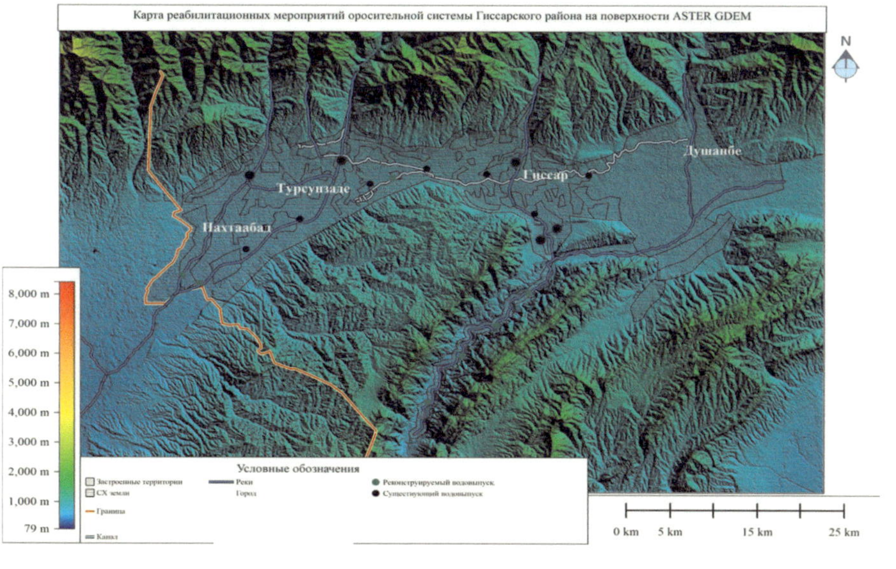

Figure 8.
Map of rehabilitation measures for the irrigation system of the Gissar Valley.

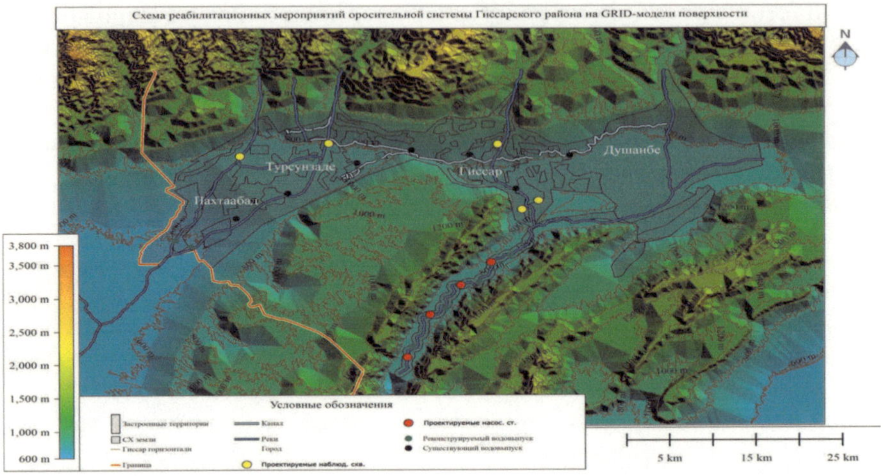

Figure 9.
Map of rehabilitation measures for the irrigation system of the Gissar Valley.

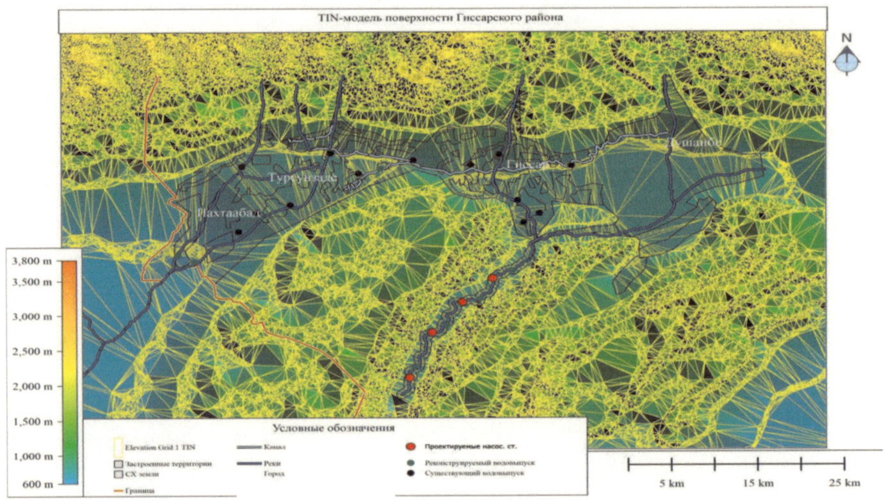

Figure 10.
Map of rehabilitation measures for the irrigation system of the Gissar Valley.

The next step is to create an irregular triangulation network based on the surface height data, a way of modeling a continuous surface with points and values at these points selected with variable density (**Figure 10**). Further, based on the TIN-model of the surface, knowing the heights, slopes and exposure of any point on the surface, a model of the drainage areas of the territory is created (**Figure 11**).

4. Summary

Obtaining relevant and timely information obtained as a result of monitoring irrigated lands is one of the main problems in solving issues of studying the state and use of irrigated lands. Formation of up-to-date and reliable information, obtained in a

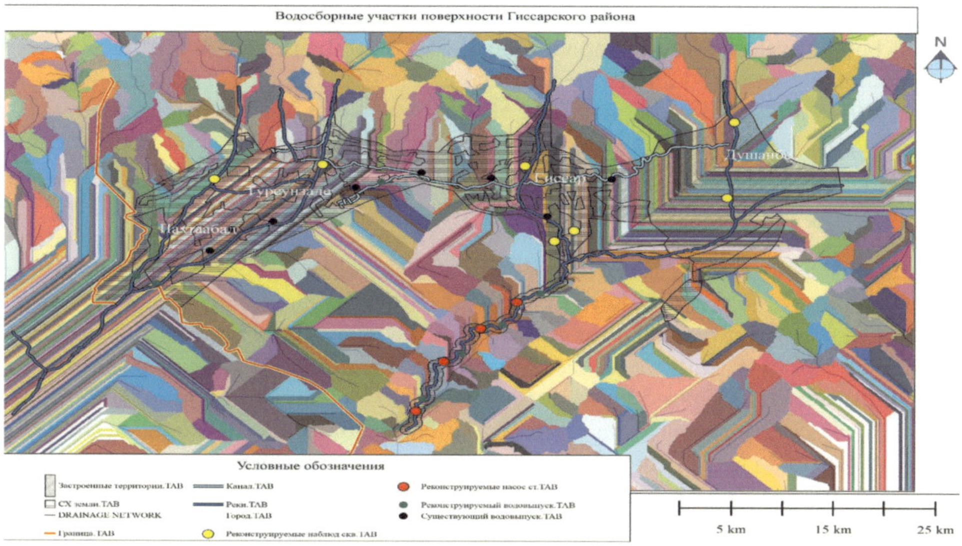

Figure 11.
Map of watershed areas of the surface of the Gissar Valley.

timely manner with the help of modern information and telecommunication technologies, and processing of the obtained large-scale big data in geoinformation systems and other software products, will allow timely assessment, design and development of a management decision for carrying out measures to improve soil fertility, protection of irrigated lands and irrigation systems.

The problem of land degradation and environmental pollution when using irrigated lands requires the improvement of theoretical and methodological approaches to monitoring irrigated lands using modern technologies for remote sensing of the Earth, obtaining big data and processing based on the use of software products for processing big data, geographic information systems also for processing and assessing the ecological state and use, which will make it possible to obtain reliable graphic information that allows it to be used for designing, predicting the state of lands and developing measures to prevent the negative impact of the impact on these lands or eliminate the consequences of these impacts.

The studies conducted by the author made it possible to obtain up-to-date and reliable information on the state and use of irrigated lands for planning the rational and efficient use of these lands. The region of the Gissar Valley of the Republic of Tajikistan was chosen for research, the analysis of the results of which allowed the author to draw the following conclusions and suggestions:

1. The study of the theoretical provisions of the monitoring of irrigated lands showed the need to collect a large amount of data that combine the characteristics of various indicators that make up the environmental objects of the study area. It has been established that the prompt and most complete obtaining of such indicators for irrigated lands is possible based on the use of modern technologies and geoinformation systems for processing the obtained data. As a result of the research, the author proposed the basic principles for building a monitoring system for irrigated lands, one of which is the principle of the mandatory formation of complex information (cadastral, ecological state of

irrigated lands and the environment, technical state of reclamation, irrigation and other systems water management). Based on the developed principles of monitoring irrigated lands, the author has developed a model for monitoring irrigated lands in the Gissar Valley of the Republic of Tajikistan, the main goal of the algorithm is to create a methodology for monitoring irrigated lands for the conditions of increased natural and anthropogenic pressure. The proposed algorithm for the monitoring model of irrigated lands in the Gissar Valley includes such stages as: study of the methodological base for monitoring lands by classification criteria; identification of natural and geographical features of the Gissar valley; determination of factors of natural and technogenic processes that have a negative impact on the ecological state of irrigated lands based on the application of various statistical programs; formation of a system of indicators for monitoring irrigated lands in the Gissar Valley; development of a methodology for monitoring irrigated lands based on the developed system of monitoring and updating indicators for study sites on the territory of irrigated lands, engineering reclamation and irrigation facilities, water management facilities, as well as the natural environment.

2. The unique location of the Republic of Tajikistan (elongation up to 700 km from west to east, with large elevation changes from 300 to 7495 m) determines the climatic and natural features, the distribution of the population on the territory of the republic, its population, and the development of economic activity. It has been established that the annual population growth averages 2.24%, the population density of Tajikistan is 65.7 people. For 1 sq. km, and in the Gissar region the density is 277 people. For 1 sq. km, which indicates the special conditions of the territory, its attractiveness for living in terms of natural and climatic conditions, as well as the possibility of developing agriculture, and this also indicates an ever-increasing anthropogenic pressure on the environment and, above all, land resources based on irrigated lands. In addition, the mountainous territories of the Republic are a habitat for unique animals, birds, mammals and rare plants listed in the Red Book.

3. The author has made an analysis of natural and technogenic factors that have a negative impact on the state of irrigated lands. The area of degraded lands in Tajikistan is about 97.9% of the territory. Every year, about 50 thousand hectares of cultivated land are subject to varying degrees of desertification. Factors that have a negative impact on irrigated lands in the studied areas of the Gissar Valley have been established, the main of which are: the impact of geomorphological, geological, climatic, soil-vegetative and economic conditions that cause soil degradation (water and wind (deflation) erosion, gully formation, waterlogging and waterlogging, salinization and schistosity, compaction and stylization). In addition to natural factors (as mentioned above), anthropogenic factors that affect irrigated lands lead to degradation, which include, for example, the state of the collector-drainage system (CDS), the length of which is 13096.2 km, the total length irrigation system is 29793.9 km.

4. It has been established that one of the main environmental pollutants in the Gissar Valley is the Aluminum plant, located in the Tursunzade district, which produces man-made pollution with emissions of heavy metals and metalloids into the atmosphere. It is shown that in the Gissar Valley, desertification is

severely affected, more than 14 thousand hectares of land, and almost 22 thousand hectares to a medium-low degree, including land in the old irrigated territory of the valley, arable land by almost two thirds are subject to water erosion (63.4%) and more than half of them are moderately and severely eroded. More than 93% of pasture lands are subject to erosion.

5. Collection and analysis of multispectral satellite images from the Landsat 7 and Landst 8 satellites using the ArcGis software was performed. In the process of their analysis, maps were obtained for the NDVI index from 2010 to 2019 and traced changes in the area of vegetation cover during 2010–2019. And then we got MNDWI maps from 2010 to 2019, water coverage, and identified areas that belong to irrigated land using the received vegetation and water coverage data.

6. Grouping of negative processes occurring in the Gissar valley into the following groups was carried out: natural, technogenic, social; and factor analysis make it possible to identify the most significant pressures on agricultural landscapes. The developed methodology for substantiating the establishment of relevant, reliable indicators, processing with the help of modern big data software products, the results of processing monitoring measurements were used to assess the state of irrigated lands in the Gissar Valley and draw up maps for the areas of the Gissar Valley, which includes: Shakhrinavsky; Tursunzadevskiy. The analysis carried out shows that the system of controlled indicators for monitoring irrigated lands fully fits into the model determined by such indicators as: state, impact, productivity.

7. Hidden factors affecting the state of irrigated lands and forecasting the development of changes in their state have been identified. The indicators for water resources and vegetation of irrigated lands, which are influenced by climatic factors, are considered: average monthly temperature; average annual precipitation; average humidity; average annual wind speed; average number of sunny days per year. The obtained indicators are processed using statistical software products JMP Statistical Discovery, Minitab 19, Statistica. First of all, the indicators are checked for the normality of the distribution of data in order to select a criterion for the purpose of further analysis. Based on the data obtained by the Holt-Winters method, using the ANOVA method, we calculate the impact of groups of indicators on irrigated land and water coverage in the next 3 years. It has been established that the following groups of factors will most strongly influence the irrigated lands: temperature and precipitation; temperature and humidity; temperature and wind speed; temperature and sun; humidity and wind speed; humidity and sun. And the water will be affected by: temperature and precipitation; temperature and humidity; temperature and wind speed; temperature and sun; precipitation and humidity; precipitation and wind speed; precipitation and sun; humidity and wind speed; humidity and sun.

8. Analysis of the qualitative state of the irrigated lands of the Gissar Valley by the author made it possible to establish that almost the entire territory of the valley is characterized by the development of processes of degradation of soil and vegetation cover, a critical ecological state. Carrying out land management work on these lands on an agrolandscape basis in combination with other works will ensure their rational use and achieve their sustainable state. For these purposes,

up-to-date information on their condition and use is required, obtained as a result of monitoring of irrigated lands based on remote sensing materials, field research and processing of a set of data obtained using various statistical software products and geographic information systems.

9. The author, based on the developed methodology for selecting the most informative indicators and using Landsat satellite images obtained for the period from 2010 to 2020, using geoinformation systems to process the received monitoring data of irrigated lands, developed maps that reflect the current state of the art. - the state of the irrigated lands of the Gissar Valley based on and developed proposals for the rehabilitation of the irrigation system of the Hissar Valley, which will ensure the sustainable development of the use of these lands.

Author details

Aliev Nozim Numonovich
Geography Science, Faculty of Land Management and Geodesy, Tajik Agrarian University named after Sh. Shotemur, Tajikistan

*Address all correspondence to: nnozim@gmail.com

References

[1] Geoanalytical system. "Agricultural management" now in the Tambov region-2013 [Electronic resource]. Available fron: http://press.scanex.ru/index.php/ru/news/item/3991-agro [Accessed: June 16, 2014]

[2] Yu KE, Yu GI. Land monitoring using GIS technologies. Scientific Journal of the Russian Research Institute of Land Reclamation Problems. 2011;**4**(04)

[3] Pronko NA, Korsak VV. Management of irrigated agriculture based on the use of information technologies. Scientific Life. 2012;**2**:80-87

[4] Pronko NA, Korsak VV, Korneva TV. GIS-monitoring of the ameliorative state of irrigated lands (on the example of the dry steppe Zavolzhye). Melioration and Water Management. 2008

[5] Bogolyubova AA. In: Bogolyubov AA, editor. Aerospace Monitoring of Lands of Protected Natural Areas of St. Petersburg: Dis. ... Cand. Tech. Sciences. St. Petersburg; 2012. p. 144

[6] Afanasiev YA. In: Afanasiev YA, Fomin SA, editors. Monitoring and Environmental Control Methods. MNEPU; 1998

[7] Baklanov AD. Ecological Zoning of Territories of Natural and Technogenic Systems of the Central Chernozem Region. In: Baklanov AD, Titova SL, Pribytkova NA, editors. Vol. 1, No. 16. Ecology of the Central Black Earth Region of the Russian Federation; 2006. pp. S. 78-S. 80

[8] Lepekhin PP, Fomin AA, Aliev NN. Problems of reclamation lands in the Gissar Valley of the Republic of Tajikistan and solutions using GIS technologies. International Journal of Applied Sciences and Technologies "Integral" Certificate of media registration in Roskomnadzor: El No. FS77-74090 No. 159-RSCI dated May 13, 2020 [Electronic resource]

[9] Danilov-Danilyan VI. In: Danilov-Danilyan VI, Zalikhanov MC, Losev KS, editors. Ecological Safety: General Principles and Russian Aspect. ProgressTradition; 2001. p. 415

Chapter 6

Smallholder Irrigation for Climate Mitigation and Cacao (*Theobroma cacao* L.) Performance Improvement in the Rainforest Tropics

Samuel Agele, Kayode Adejobi and Abel Ogunleye

Abstract

Climate change poses significant threats to agriculture, including food security, livelihoods and economic growth. Based on the importance of cocoa, there is a need for sustainable crop production and resilience to anticipated changes in rainfall and temperature in the future. Irrigation is an important climate-smart practice for alleviating abiotic stress and enhancing crop productivity, and irrigation is seldom practiced in the cacao orchards of West Africa. Studies were conducted to examine the effects of dry season gravity drip irrigation on the rootzone moisture, tree water use (evapotranspiration), leaf area index and yield of cacao in a rainforest zone of Nigeria. Irrigation treatments were based on water application at 5- and 10-day intervals and 50, 70 and 100% Pan evaporation, which was applied using point source emitters on drip lines. The soil moisture content, photosynthetic active radiation, leaf area index and extinction coefficient differed among the irrigation treatments. Deficit irrigation (10-day and 50% EPan) enhanced water use efficiency by 25–44% (30 and 50% water savings), while full irrigation enhanced soil moisture, cacao ET, and pod and bean yields. This study established irrigation and water requirements for cacao in the dry season and confirmed the relevance of irrigation for enhanced cacao performance and climate mitigation.

Keywords: *Theobroma cacao*, seasonal transitions, rainforest, irrigation, performance, climate stress

1. Introduction

Cocoa (*Theobroma cacao* L.) is an important perennial fruit tree with an estimated annual global production of 3.2 million tonnes [1]. In West African cocoa-producing countries, cocoa is a major foreign exchange earner and provides employment for millions of smallholder farmers whose small farm sizes range from 0.5 to 5.0 ha. In Nigeria, the main cocoa-producing areas are in the rainforest south where an estimated.

A total of 1.45 million hectares are cultivated, and cocoa productivity is low (approximately 250 kg dry beans/ha) compared with 600–1000 kg/ha for Cote d'Ivoire and Indonesia.

Cocoa is the most prominent perennial fruit tree species in the rainforest of Nigeria. Fruit trees are characterized by deciduous growth habits but are cultivated under rainfed conditions [2, 3]. The annual rainfall in the cocoa growing region of West Africa has a bimodal distribution of less than 2000 mm, resulting in wet–dry seasonal transitions [2, 4]. The dry season lasts approximately 3–4 months and is characterized by dry and hot weather. Smallholder farmers rarely practice irrigation; nevertheless, irrigation is effective for meeting crop water demand and ameliorating climate-related stresses [5, 6]. In West Africa, cocoa is cultivated as a rainfed crop and is thus subjected to poor soil and weather conditions [3, 5, 7]. Cocoa is cultivated as a rainfed crop, and it is highly sensitive to soil and weather conditions such as low rainfall, soil and air moisture deficits and temperature stresses [3, 5, 7]. The changing environmental conditions (marginal soils and extreme weather events) impose constraints on cacao growth and productivity. The annual rainfall in the cocoa growing region of West Africa follows a bimodal rainfall distribution pattern of less than 2000 mm (wet–dry seasonal transitions: [2, 4]). During the 3 months of dry and hot weather, smallholder farmers rarely practice irrigation. However, irrigation is effective for meeting crop water demands and ameliorating climate-related stresses [6].

Climate change, extreme and variable weather conditions and other climate-related disasters, such as high frequency and severe drought events and warming temperatures (1.5–2°C), occur worldwide [8]. These changes may result in changes in regional precipitation and evapotranspiration [9, 10]. Among natural hazards, drought ranks first in terms of the number of people directly affected [11].

Climate mitigation may be built on adaptation practices to strengthen the resilience of farming systems to anticipated changes in rainfall patterns and temperature in the future. Climate-smart cocoa production landscapes may be built on agricultural innovations such as high-yielding drought-tolerant crop varieties, climate information services, agricultural insurance, nutrient and water management via irrigation, mulching, etc. Sustainable management practices for cacao (*Theobroma cacao* L.) in a changing climate may include irrigation schemes for climate stress alleviation and enhanced productivity.

The available information on cocoa water use in the field showed estimated values ranging from 3 to 6 mm/day during rains and less than 2 mm/day in the dry season [12, 13]. The FAO Penman–Monteith equation is accepted worldwide as the standard method for estimating reference evapotranspiration (ETo). The ETo indicates crop consumptive water use as the sum of evaporation from soil and plant transpiration f [13–16]. FAO Paper No. 56 highlights procedures to calculate ETo from radiation, wind, humidity and temperature data in addition to the crop coefficient K_c for crops [17]. Allen et al. [17] suggested a K_c value of 1.0–1.05 for a cocoa crop with a complete canopy. Field reports of cocoa water use (ETc) have shown values ranging from 3 to 6 mm/day during rains and less than 2 mm/day in the dry season [12, 13], while data based on the sap flow method suggest values less than 2 mm/day. Moser et al. [18] conducted a simulated El Niňo drought experiment and reported no significant differences between a rainfed control treatment and rain through-fall reduction (70–80% under a dry soil profile near the permanent wilting point). The reports showed that the maximum cacao bean yields were obtained for cocoa drip irrigated with 20 l/tree/day (175 l/tree as total irrigation). Based on field trials, Diczbalis et al. [19] reported an annual irrigation requirement of 470 mm for cacao (a maximum weekly requirement of approximately 200 l/tree (1250 trees/ha)) and dry bean yields between 1.5 and 2.7 t/ha for young fruiting trees.

Information on soil moisture extraction, water use and cocoa yields under dry irrigation from the rainforest belt of West Africa is inadequate. Thus, experiments were designed to examine the effects of regulated dry season irrigation on the root zone moisture, tree water use and bean yield of cacao in a rainforest zone in Nigeria.

2. Materials and methods

2.1 Experimental site and conditions

Experiments were conducted in the field using 5-year-old cacao trees that had been previously irrigated since the first year of field establishment (2012). The study was carried out at the Research Station of the Federal University of Technology, Akure, in the rainforest zone of Nigeria. In agroecology, rainfall patterns are characterized by bimodal and seasonal wet–dry transitions. The dry season is characterized by terminal drought caused by inadequate rainfall, soil moisture deficits, high vapor pressure deficits and temperatures and very clear skies [20].

2.1.1 Irrigation strategies

The drip irrigation system (drip irrigation) was laid out in the field. This included a pumping machine, good water source, pipes, drip lines, overhead tank (with stand), and pressure control valves. Water was applied via a gravity-drip irrigation system via point source emitters, which were installed on the laterals of each row of crops. The emitters were installed on the laterals of each row of crops and were spaced 3 m apart. The irrigation buckets were suspended on 5 m high tank stands to provide the required hydraulic heads [6]. The low-head (gravity) drip system supplied water to plant roots via drippers using inline emitters with a discharge rate of 2 l/h, which were spaced at 3 m intervals laterally. One drip lateral served each plant row. An inflow meter was installed at the control unit to measure the total flow distributed to all replications in each treatment.

2.2 Experiment 1: irrigation treatments at 5- and 10-day intervals

Preliminary studies based on variable irrigation amounts and frequencies for cacao in the study area have shown promising results (Agele, personal communication) [2]. The present study is a follow-up study aimed at validating the split-application of 14.28 mm (3.86 l/day) at 5- and 10-day irrigation intervals in the field. A pretreatment of 135 mm of irrigation water was applied to replenish the soil water within a 0.60 m profile depth (to field capacity). Irrigation was applied at 5- and 10-day intervals and was arranged in a split-plot design.

Irrigation was imposed based on the restoration of cumulative potential evapotranspiration (ETo) via the FAO method [17, 21] in the following form:

$$\mathrm{ETa} = K_c \mathrm{ETo} \tag{1}$$

where ETo is the potential evapotranspiration and K_c is the crop coefficient [17, 21].

A crop coefficient (K_c) of 0.83 was adopted for cacao (which was in the early fruiting stage) [17].

The potential evapotranspiration (ETo) from December to May was derived from the Penman–Monteith combination equation [17, 21] using data obtained from the Meteorological Observatory of the Experimental Station.

The water requirement (WR) was determined using the following relation:

$$\mathrm{WR} = A * B * C * D * E \quad (2)$$

where WR = water requirement (l day/plant), A = open Pan evaporation (mm/day), B = Pan factor (1.0), C = spacing of plant (m^2), D = crop factor. The crop factor depends on plant growth; the value for fully grown cacao was 1.13, but for cacao in the early fruiting stage, 0.83 was adopted.

The irrigation amount (volume per application) was calculated as follows:

$$V = P * A * E_{\mathrm{Pan}} * K_{\mathrm{cp}} \quad (3)$$

where V is the volume of irrigation water (l); P is the wetting percentage (taken as 100% for row crops); A is the plot area (m^2); and E_{Pan} is the pan evaporation and K_{cp} pan coefficient (1.0). This corresponded to 14.28 mm (3.86 l/day), an amount that was applied at 5- and 10-day intervals.

The irrigation WR was determined using seasonal Pan evaporation data for the area. The total water requirement (TWR) of the farm plot was obtained as follows:

$$TWR = WR * Number\ of\ plants \quad (4)$$

where TWR is the total water requirement and WR is the water requirement (l day/plant).

The maximum allowable deficit (MAD) for cacao was assumed to be 50% of the available water storage capacity of the soil (AWC).

The actual evapotranspiration (consumptive water use: ETc) of cacao trees was derived from the water balance equation (Eq. (1)) [6].

$$\mathrm{ET} = I + P + \mathrm{dS} - \mathrm{Dp} - \mathrm{Rf} \quad (5)$$

where ET is the actual crop evapotranspiration (mm) and I is the amount of irrigation.

Water applied (mm); P, precipitation (mm); dS, change in the soil water content (mm); Dp, deep percolation (mm); Rf, amount of runoff (mm). Since the amount of irrigation water was controlled, deep percolation and runoff were assumed to be negligible.

Soil water measurements were taken throughout the growing season using the gravimetric method.

The maximum (management) allowable deficit (MAD) for cacao was set at 50%.

Cacao LAI and solar radiation integrals (incident, transmitted and absorbed radiation, the ratio of radiation measurements below and above the canopy and PAR) were measured using a canopy analyzer (Delta T, UK). The incident solar radiation (R_I) above the canopy was measured using a pyranometer connected to the canopy analyzer system.

The line sensor was attached to a metal frame and lifted above the cacao canopy.

Photosynthetic active radiation (PAR) was measured in addition to solar radiation. The analyzer measures light transmitted by the ratio of radiative measurements below and above the canopy [20]. The fraction of intercepted radiation (R_{fract}, %) was calculated as follows:

$$R_{\text{frcat}} = (R1 - R_{\text{T}}/R_{\text{I}} \tag{6}$$

where I is the incident radiation above the canopy and T is the transmitted radiation.

Canopy extinction coefficient.

The Beer–Lambert law describes the absorption of light by plant pigments in solution. This function demonstrates that the absorption of light will be more or less exponential with increasing intercepting area down through the canopy. The light extinction coefficient (k), according to the Beer–Lambert law, is:

$$k = \left[\text{loge}\left(\frac{I}{Io}\right)\right]/\text{LAI} \tag{7}$$

$$k = -\ln\ (I - I_{\text{o}})/\text{LAI} \tag{8}$$

where I and I_{o} are the irradiance values upon and under the canopy, respectively.

LAI is the LAI of leaves causing light attenuation, and k is the extinction coefficient or slope of the curve when the natural log (In) I/I_{o} is plotted against the LAI. The light extinction coefficient (k) was calculated by inverting Lambert–Beer's law Taku et al. [22]:

$$K\text{df} = -\ln\ (0.94\ \text{PAR transmitted}/\text{LAI} \tag{9}$$

Fractional radiation (I) interception was calculated according to the following equation:

$$I = (R_{\text{i}} - R_{\text{t}})/R_i \tag{10}$$

where R_{i} is the incident radiation and R_{t} is the transmitted radiation.

The proportion of transmission (TR) from the incident radiation (Ra) was obtained by the following formula:

$$\text{TR}(\%) = \left(\frac{\text{Rb}}{\text{Ra}}\right) * 100 \tag{11}$$

2.3 Experiment 2: irrigation treatments with 50, 70 and 100% pan coefficients (K_{cp})

The irrigation treatments were based on variable Pan coefficients and were delivered using point source emitters on gravity-powered drip lines installed on rows of trees [6, 17, 23]. Pan coefficients (100, 70 and 50% K_{cp} amount to 0, 0.3 and 0.5 relative water deficits, respectively). Thus, the amount of irrigation water to be delivered was derived from the product of Pan evaporation and Pan coefficients (K_{cp}) [17, 21]. The adopted coefficients were 1, 0.7 and 0.5, respectively:

$$\text{Ir} = A^*\ E_{\text{Pan}}{}^*\ K_{\text{cp}} \tag{12}$$

where Ir is the amount of applied irrigation water (mm), A is the plot area, and E_{Pan} is the cumulative evaporation at the irrigation interval (mm) and K_{cp} are the plant-pan coefficients. The irrigation treatments were IrT1 (E_{Pan} * 100 K_{cp}) and IrT2 (E_{Pan} * 70% K_{cp}) and IrT3 (E_{Pan} * 50% K_{cp}).

These treatments denote the adequacy of water delivery (IrT1: the noncrop water stress baseline), and IrT3 denotes the maximum water deficit (the stressed baseline).

The actual crop evapotranspiration (ETc) of the cacao trees was calculated according to Eq. (5), and the TWR was determined according to Eq. (2) (Experiment 1). Therefore, the irrigation requirements (WRs) were 9.63, 6.75 and 4.82 l/plant/day for the IrT1, IrT2 and IrT3 irrigation treatments, respectively. The seasonal irrigation amount ranged from 4.82 l/day and 127,500 mm (at the DI_1 level) to a minimum of 1.93 l/day and 20,400 mm (at the DI_4 level). The irrigation WR and TWR of the farm plot were obtained according to Eq. (4) (Experiment 1).

The gross irrigation requirement (GIR) of the coca field (computed as ETc) is considered the net irrigation requirement (NIR), which is obtained following its division by the application efficiency (AE).

$$\text{GIR} = \text{NWR} = \text{AE} : \tag{13}$$

2.4 Orchard water use efficiencies

Crop water productivity (efficiencies of crop water use WUE) and irrigation treatments (irrigation use efficiency: IWUE) were determined using the methods of Sezen et al. [23] and Agele et al. [6]:

$$\text{IWUE} = \text{biomass weight } (Y)/\text{total irrigation water applied (Ir)} \tag{14}$$

$$\text{WUE} = \text{biomass weight } (Y)/\text{cumulative seasonal ETc} \tag{15}$$

where IWUE is the irrigation water use efficiency (t/ha/mm) and EY is the economic yield (t/ha), and Ir is the amount of applied irrigation water (mm).

3. Results and discussion

3.1 Effects of 5- and 10-day irrigation intervals on cocoa production

Compared with the dry season (December to March), the rainy season (March to early December) had a higher mean relative humidity average (71%), high cloud overcast (overcast sky) and lower air temperatures (32.8°C). The cumulative amounts of seasonal irrigation water delivered were 12,119, 8483 and 6059.3 mm (**Figure 1**). Compared with the 10-day interval, the 5-day irrigation treatment resulted in greater water delivery to the cocoa root zone across the sampling dates. The soil moisture contents differed among the irrigation water deliveries and measurement dates, and higher soil moisture contents were obtained for the 5-day irrigation interval than for the 10-day irrigation interval (**Figure 2**). On average, the soil moisture content was 48% greater for the 10-day irrigation treatments (**Figure 3**) than for the 5-day irrigation treatments. The lowest soil moisture contents were obtained for DOY 45 and 120, and the highest soil moisture contents were found between DOY 345 and 75 and between DOY 120 and 150 (**Figure 2**). There were significant differences ($P < 0.05$) in soil moisture between the 5- and 10-day irrigation intervals, while the 5-day

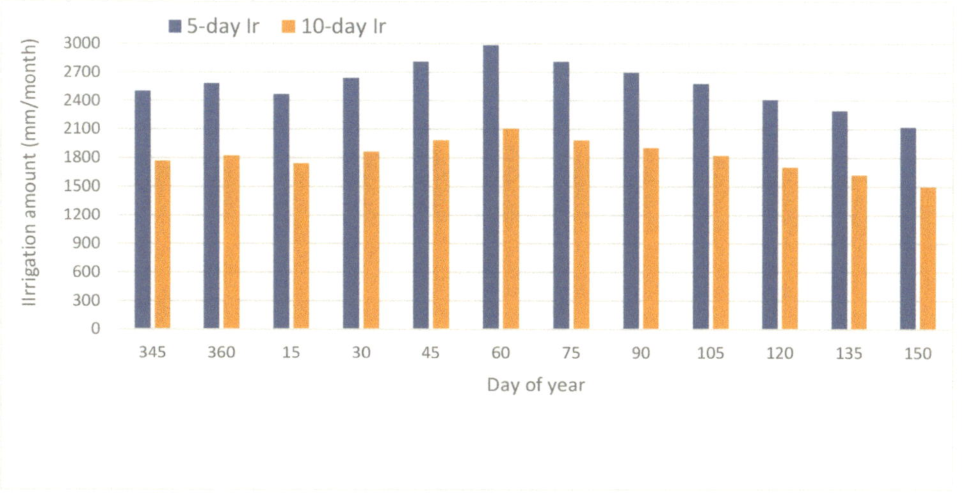

Figure 1.
Seasonal irrigation water was applied (at 5- and 10-day irrigation intervals).

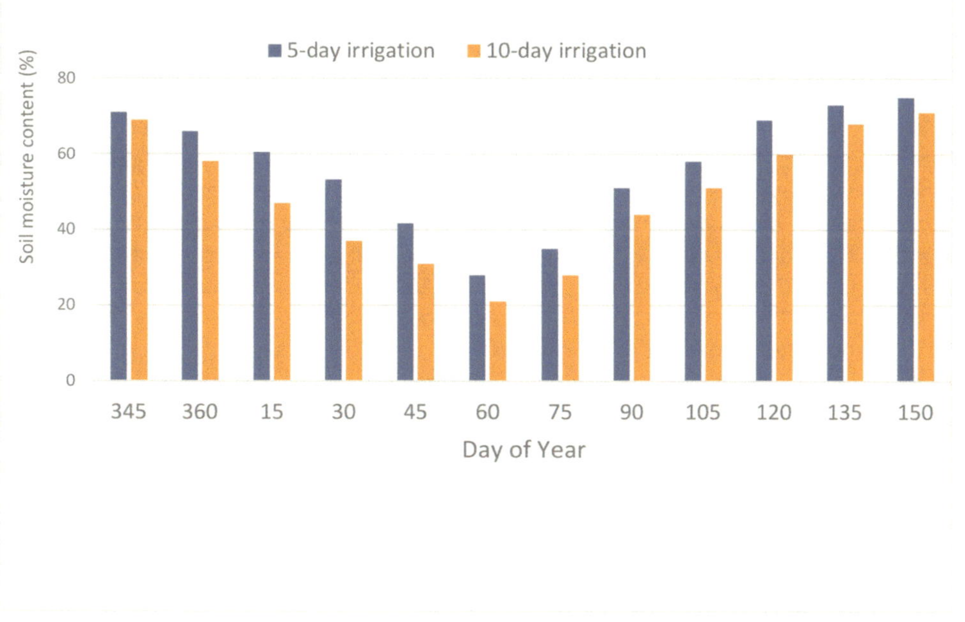

Figure 2.
Effect of irrigation on the soil moisture status.

intervals of irrigation delivered more irrigation and enhanced the soil moisture status compared with the 10-day intervals. The time course of the status of soil water before and after irrigation using moisture content measurements from soil samples within the 0–20 cm soil profile depth before and 1 day after each irrigation. At times, both irrigation intervals had values close to the wilting point between irrigation events, under which available water fell below 50% more often than not during the period of study. With more frequent irrigation (i.e., 5 days of irrigation), the soil moisture was

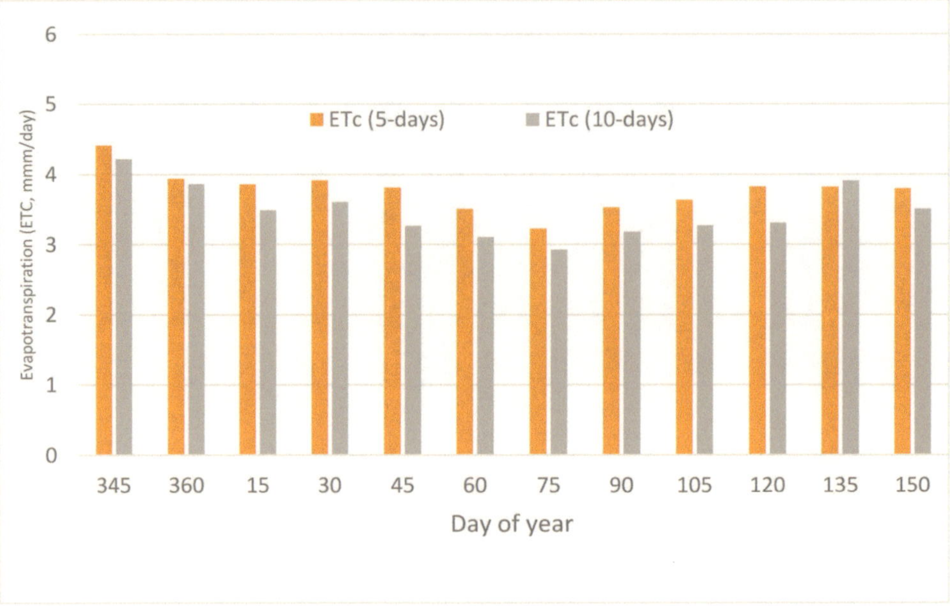

Figure 3.
Cacao evapotranspiration (ETc) as affected by irrigation intervals.

mostly within the field capacity range. In general, based on the values of soil moisture, the stored water within the crop rootzone profile was used between irrigation cycles.

Soil moisture depletion over two measurement days was analyzed to determine the soil water use (ETc). Cacao water use (ETc) differed across measurement dates and irrigation treatments (**Figure 3**). The mean calculated evapotranspiration (ETc) values were 3.72 and 3.54 mm/day, 3.44 and 3.15 mm/day, and the seasonal totals were 48.2 and 38.3 for the 5- and 10-day irrigation intervals, respectively. The average Cacao evapotranspiration (ETc) for the 10-day irrigation treatment was 45% less than that for the 5-day irrigation treatment. The time course of cacao evapotranspiration (ETc) showed that cacao water use declined between DOY 345 and 45, while the lowest values were found for DOY 45 to 105.

for both 5- and 10-day irrigation intervals (**Figure 3**). Pod and bean yields were significantly different under the 5-day irrigation treatment, which produced more and heavier pods and beans than under the 10-day irrigation treatment. The pod and bean weights for the 5- and 10-day irrigation treatments were 78,000–6000 kg/plant and 4.8–3.2 t/ha, respectively, and the water productivities were 0.45–0.33 mm/kg/ha (irrigation efficiencies) and 0.11–0.09 mm/kg/ha (crop water use efficiencies), respectively (**Table 1**).

Solar radiation integrals (the ratio of transmitted to incident radiation, PAR and LAI), the cacao canopy cover (LAI) and PAR intensities and the canopy extinction coefficient (k)) were greater under the 5-day irrigation interval than under the 10-day irrigation interval. AnimKwampong and Frimpong [4] and Agele et al. [20] reported that canopy size and resultant shade intensities affect light characteristics within cacao fields, particularly the ratio of transmitted to incident radiation, PAR and LAI [2, 24]. Anim-Kwampong and Frimpong [4] suggested LA1 > 1 as the threshold canopy cover of the land surface for optimum light interception and transmission. A relatively high LA1 will result in mutual shading, limiting light transmission and photosynthetic

Irrigation	Seasonal irrigation	Soil moisture content (%)	ETc	Seasonal ETc	Pod weight/plant (kg)	Bean weight/ Plant (g)	Bean yield (ha)	WUE (bean) (Irr)	WUE (bean) (ETc)	PAR	LAI	K
5-Days	12,452	19.2	3.71	43.85	7478	816.6	4.59	0.043	0.101	628.1	3.51	0.61
10-Days	6226.2	16.4	3.23	41.44	6352	708.3	3.53	0.035	0.083	585.7	3.14	0.57
LSD (0.05)		1.9	0.81	1.14	0.94	115.4	1.06	0.007	0.003	61.3	0.31	0.03

Table 1.
Summary of measured soil and cacao parameters as affected by 5- and 10-day irrigation intervals.

activity. The Cacao LAI affects light attenuation and other radiation characteristics within the canopy, such as an increase in diffuse light [25]. Diffuse radiation is associated with increases in canopy light attenuation (canopy extinction coefficient). The canopy extinction coefficient (k) was greater for the 5 days of irrigation, which also had a greater LAI. The higher extinction coefficient for the 5-day irrigated cacao may be due to the lower amount of transmitted light (diffuse radiation). Goudriaan and Monteith [26] affirmed that a low extinction coefficient enhanced growth when the plant canopy was fully developed and uniformly distributed. Acheampong et al. [27] reported that biomass accumulation and overall development of cacao depend on the intensity of the PAR received. Irrigation ameliorated the microclimate via a reduction in thermal load and improved the soil moisture status and crop evapotranspiration [2, 18]. Such modification of the microclimate would enhance CO_2 assimilation by leaves and Charles et al. [2] reported that a large extent of tree canopy will increase vegetative growth and photosynthetic activity of leaves. Reduced hydrothermal stress has implications for cacao survival and productivity during the terminal drought situation of the dry season in the study area.

Cacao irrigation and WRs have been variously studied [12, 13, 18, 19, 28]. Cacao water use values were reported to range from 1.3, 1.15, 3 and 5 mm/day using E_{Pan} coefficients between 1.0 and 0.6 and 224 mm cumulative seasonal water use during the dry season and weekly irrigation requirements of 470 mm and 200 l/tree. Penman [13] computed 3–5 mm/day as an ETc and an ETc of approximately 1.3 mm (10 l/tree/day), Kohlerlscher et al. [28] reported 2 mm/day, and Moser et al. [18] reported 1.3–1.5 mm/day. In Cote d'Ivoire, the depth of irrigation of 920–1650 mm and E_{Pan} coefficients between 1.0 and 0.6 and 224 mm cumulative cacao water use during the dry season produced between 30 and 60% of the cacao bean yield increase [19]. The authors applied 470 mm of seasonal irrigation and 200 l/tree for weekly irrigation, resulting in bean yields of 1.5–2.7 t/ha. The present obtained cacao yields ranged between 7800 and 6000 kg pods/plant and between 430 and 280 g beans/plant (4.8 to 3.2 t/ha). The more frequent irrigation out-yielded deficits of 20 and 24% for pods and beans, respectively. More frequent irrigation (at 5-day intervals) than at 10-day intervals enhanced the soil moisture content and water use (ETc).

3.2 Effects of irrigation using pan coefficients (Kcp) of 1.0, 0.7 and 0.5 on cocoa production

The irrigation regimes computed using 100, 70 and 50% Pan coefficients affected the soil moisture content, water use, pod and bean yields and water productivity (irrigation and crop evapotranspiration) of cocoa. Irrigation with Pan coefficients (Kcp) of 1.0, 0.7 and 0.5 delivered different amounts of water to the cacao rootzone. The irrigation amounts (monthly averages) were 1009.88, 706.91 and 504.94 mm, while the seasonal totals were 2116.5, 8482.95 and 6059.25 mm for the IrT1, IrT2 and IrT3 treatments, respectively (**Figure 4**). Irrigation at IrT1 (E_{Pan} * 100 K_{cp}) had the greatest effect, and irrigation at IrT3 (E_{Pan} * 50% K_{cp}:0.5) had the least effect. Compared with IrT1, IrT2 and IrT3 delivered 79 and 68% more water to the cacao rootzone, respectively. The maximum irrigation amount occurred on DOY 45, 60, 75 and 90, which coincided with the highest E_{Pan} values (>5 mm/day).

Irrigation treatments (IrT1, IrT2 and IrT3) affected the soil moisture content within the cacao root zone. (**Figure 5**). The highest soil moisture content was obtained for the well-irrigated treatment (IrT1), and the lowest was obtained for the deficit irrigation treatment (IrT3). For the respective deficit irrigation treatments (IrT2 and

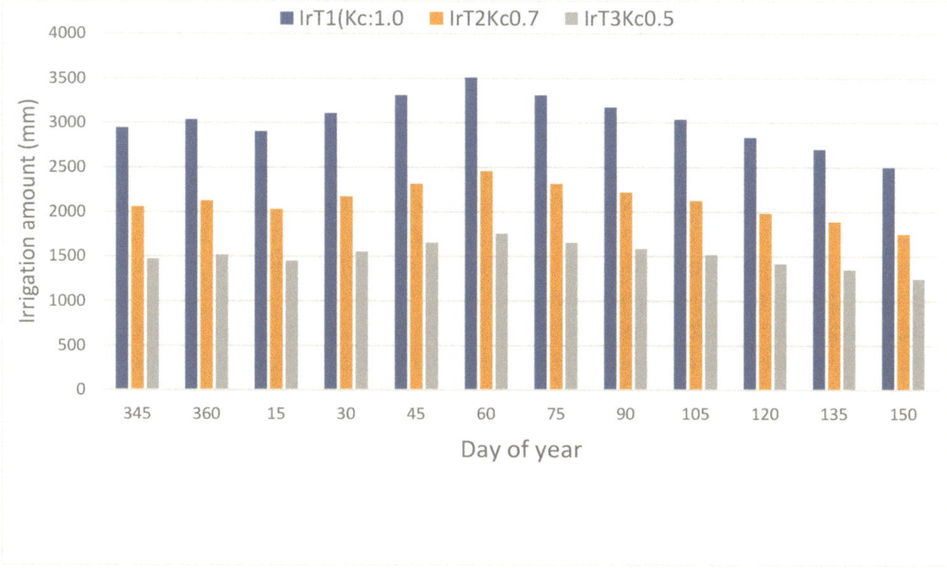

Figure 4.
Seasonal irrigation delivery (IrT1, IrT2 and IrT3).

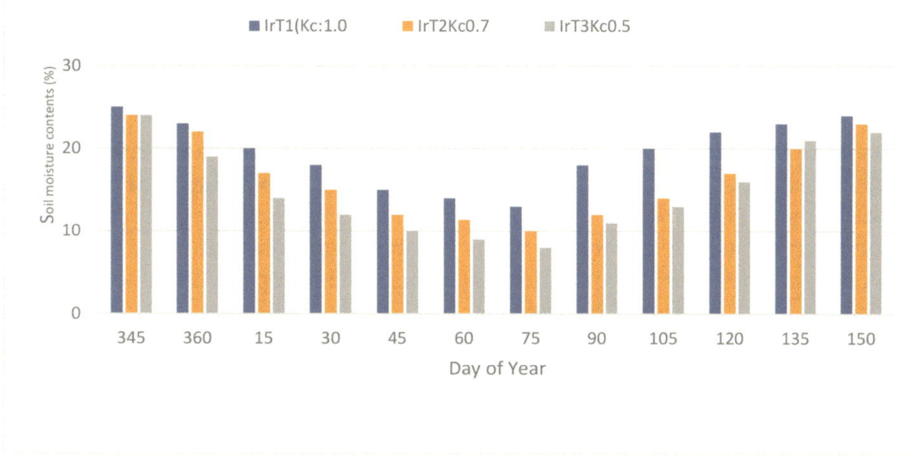

Figure 5.
Soil moisture contents as affected by irrigation treatments.

IrT3: 0.7 and 0.5 Pan coefficients), the average soil moisture contents were 61, 48 and 42% lower than those in IrT1, respectively, and irrigation equated to 30 and 50% water savings (**Figure 5**). The decreasing trends of the calculated evapotranspiration (ETc) were IrT1 (9.6 l/tree/day) > IrT2 (6.8 l/tree/day) > IrT3 (4.8 l/tree/day).

Decreases in soil moisture contents were obtained from DOY 345 to 60, followed by increasing trends in soil moisture from DOY 75 until the end of the measurement (DOY 150). The decreasing trends in the values of soil moisture content may be attributed to the increasing intensities of climatic demand (high vapor pressure deficits).

Unfavorable weather, high temperatures, soil evaporation and low atmospheric humidity enhance soil moisture depletion and thus decrease the soil moisture status

[5, 29]. In general, the observed trends in the status of rootzone moisture are attributable to the prevailing weather conditions, which are denoted by increasing intensities of climatic demand (vpd) and temperatures during periods (DOY 345 to 60) of the experiment. An increase in moisture was observed from DOY 75 until the end of the experiment (DOY 150) can be attributed to rainfall received following its commencement (mid-March). In general, the soil moisture within the crop rootzone profile decreased between irrigation cycles. This may be attributed to the intensities of climatic stress (high temperatures and vapor pressure deficits), which presumably enhanced soil evaporation. The soil moisture reserve was unable to meet the cacao water demand during the dry season [2, 30].

Cacao water use (ETc) was determined from two soil moisture measurement cycles. Cacao water use differed across measurement dates and irrigation levels (**Figure 6**). The average calculated evapotranspiration (ETc) values were 139, 97 and 63 mm/day for the IrT1 (K_c:1.0), IrT2 (K_c 0.7) and IrT3 (K_c 0.5). Cacao evapotranspiration (ETc) under deficit irrigation (IrT2 and IrT3) was 45 and 70% lower than that under IrT1 (100 K_{cp}). The observed differences in ETc values during the experimental periods are attributable to changes in weather conditions. Following the commencement of rainfall and associated replenishment of soil moisture, lowering of temperatures (air and soil) and high atmospheric humidity (declining atmospheric demand), high amounts of cacao water were obtained.

Yang et al. [31] recorded higher soil moisture status and cacao ETc for irrigated citrus plants using full-pan evaporation. The magnitude of the calculated ETc was also within the range reported by Carr [12]. Penman [13] obtained 3–5 mm/day during rains and less than 2 mm/day in the dry season for cacao under an irrigation regime of 10 l/tree/day. Similarly, Kohlerlscher et al. [28] reported an ETc of 2 mm/day, while Moser et al. [18] reported an ETc of 1.3–1.5 mm/day. A field study using sap flow sensors reported 2 mm/day for cacao water use, which is lower than that in earlier reports. For example, Penman [13] reported a potential ETo estimate of 3–5 mm/day using the Penman equation. Irrigation replenished soil moisture depletion, while cacao shade offered soil surface cover, creating a favorable microclimate for the trees and consequently reducing soil evaporation [9]. Studies have reported that the conditions at the soil surface are affected by wetting via irrigation (amount and irrigation intervals) in addition to soil exposure to light [31–34]. Soil surface conditions are known to

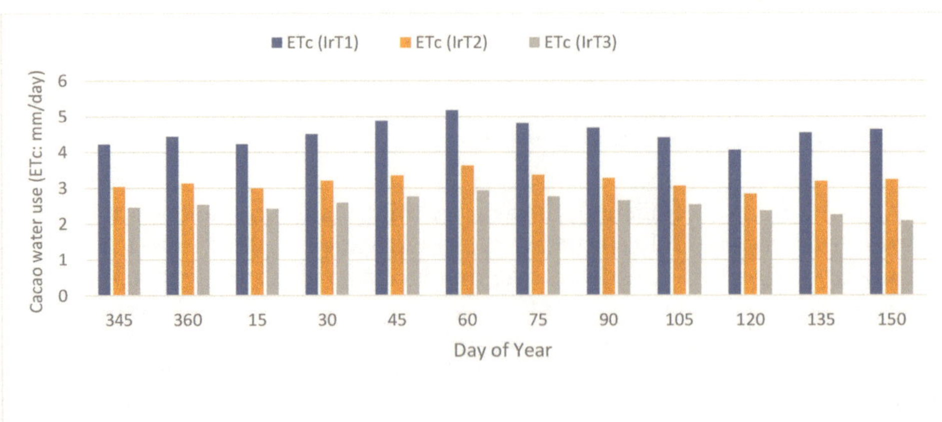

Figure 6.
Cacao water use (evapotranspiration; ETc).

DOI: http://dx.doi.org/10.5772/intechopen.112674

Irrigation	Seasonal irrigation	Scm (%)	ETc single (kc:1.13)	ETc Dual (kr t: 1.04)	ETc/E_{Pan} ratio	Seasonal ETc	Pod weight/ plant (kg)	Bean weight/ plant (g)	Bean yield (kg/ha)	WUE (Irr)	WUE (ETc)
IrT1	33858.2	21.4	5.07	5.2	0.92	139.1	4429	396.5	4.41	0.0117	0.032
IrT2	32705.3	17.3	3.55	3.7	0.73	97.3	3125	334.3	3.72	0.0142	0.043
IrT3	16929.4	14.4	2.63	2.8	0.56	62.7	2673	308.1	3.42	0.0182	0.055
LSD (0.05)		4.1	1.8	1.6	0.21	17.3	112.3	23.8	0.25	0.003	0.005

IrT1 (irrigation @ 100% K_{cp}), IrT2 (irrigation @ 70% K_{cp}) and IrT3 (irrigation @ 50% K_{cp}).

Table 2.
Summary of measured soil and cacao parameters as affected by irrigation at 100, 70 and 50% pan coefficient (K_{cp}).

determine the magnitude of crop water use (evapotranspiration: ETc) in orchards [9, 29].

The irrigation regime affected the pod and bean yields of the cacao plants. The values were significantly greater for IrT1 (35.4 and 2.29 t/ha) than for IrT2 (22.1 and 1.37 t/ha) and IrT3 (10.3 and 1.03 t/ha); thus, bean yields decreased by 60 and 40%, respectively, under IrT3 and IrT2 (**Table 2**). Carr [12] and Charles et al. [2] reported that in addition to irrigation, the yield of cacao also depends on soil properties such as infiltration rate and water holding capacity. Other studies have reported the effects of irrigation on the biomass, pod and bean yields of cacao. Diczbalis et al. [19] applied seasonal irrigation of 470 mm weekly at 200 l/tree and obtained bean yields of 1.5–2.7 t/ha.

The effects of the irrigation regime on the water productivity of cacao were evaluated. The ratio of yield to evapotranspiration (*Y*/ETc) ranged between 0.3 and 0.04 t/mm, while the yield to irrigation amount (*Y*/Irrig) ranged between 0.16 and 0.19 kg/mm (**Table 2**).

This confirmed the superiority of deficit irrigation in terms of water savings (30 and 50%) over the well-watered treatment.

4. Conclusions

Dry season irrigation enhanced soil moisture status, tree water use, canopy characteristics, pod and bean yield of cacao in a rainforest zone of Nigeria. Solar radiation properties (transmission through the cacao canopy, photosynthetic active radiation (PAR)) and canopy area and light attenuation (extinction coefficient, *k*) differed among the irrigation treatments. Compared with the 10-day irrigation interval, the 5-day irrigation interval enhanced the soil moisture status, cacao water use (ETc), and pod and bean yields of cacao. The deficit irrigation treatments (10-day intervals, especially irrigation at 50% Pan coefficient) increased the water use efficiency (25–44%), which translated to 30 and 50% water savings, respectively.

Smallholder gravity drip irrigation can ameliorate climate stress and enhance cocoa performance and is recommended for scaling-up. This study established irrigation and WRs using variable irrigation intervals and Pan coefficients for cacao during the dry season in the rainforest zone of Nigeria. These findings will inform water management decisions for optimizing the yield and water productivity of cacao, especially in the era of warming and drought. The water-saving advantage of deficit irrigation strategies can be scaled up for adoption.

Author details

Samuel Agele*, Kayode Adejobi and Abel Ogunleye
Plant Physiology and Ecology Group, Department of Crop, Soil and Pest Management, Federal University of Technology, Akure, Nigeria

*Address all correspondence to: soagele@futa.edu.ng

References

[1] Food and Agriculture Organization (FAOCLIM). Environment and natural resources service. In: Working Papers 5. Rome: FAO; 2012

[2] Charles EF, Agele SO, Aiyelari OP, Famuwagun IB, Faboade E. Shade and irrigation effects on growth, flowering, pod yields and cacao tree survival following 5 years of continuous dry season irrigation. International Journal of Environment and Climate Change. 2019; **10**(7):54-64. DOI: 10.9734/ijecc/2020/v10i730211

[3] Opeke LK. Tropical Commodity Crops. Institute, Nigeria: Spectrum Books Ltd.; 2006. p. 213

[4] Anim-Kwapong GJ, Frimpong EB. Vulnerability of agriculture to climate change: Impact of climate change on cocoa production. In: Report of Vulnerability Assessment, the Netherland's Climate Change Studies Assistance Progarmme Phase 2. Tafo, Ghana: Cocoa Research Institute of Ghana; 2004

[5] Agele S. Global warming and drought, agriculture, water resources, and food security: Impacts and responses from the tropics. In: Leal Filho W, Luetz J, Ayal D, editors. Handbook of Climate Change Management. Cham: Springer; 2021. DOI: 10.1007/978-3-030-22759-3_183-1

[6] Agele SO, Anifowose AY, Agbona AI. Effects of irrigation scheduling on components of water balance and performance of dry season fadamagrown pepper in an inland-valley ecosystem in a humid tropical environment. International Journal of Plant and Soil Science. 2014;**4**(2):171-184

[7] Zuidema PA, Leffelaar PA, Gerritsma W, Mommer L, Niels PR, Anten NPR. A physiological production model for cocoa (*Theobroma cacao*): Model presentation, validation and application. Agricultural Systems. 2005; **84**:195-225

[8] Intergovernmental Panel on Climate Change (IPCC). Climate change the IPCC's fifth assessment report. What's in it for Africa? The physical science basis. In: Headline Statements & Summary for Policymakers. 2014. 35 p

[9] Tombesia S, Frionia T, Ponia S, Palliottib A. Effect of water stress "memory" on plant behavior during subsequent drought stress. Environmental and Experimental Botany. 2018;**150**:106-114

[10] Sheffield J, Wood EF, Michael L, Roderick ML. Little change in global drought over the past 60 years. Nature. 2012;**491**:435-438

[11] Trenberth KE, Dai A, van der Schrier G, Jones PD, Barichivich J, Briffa KR, et al. Global warming and changes in drought. Nature Climate Change. 2014;**4**:17-22

[12] Carr MKV. The water relations and irrigation requirements of cocoa (Theobroma cacao L.): A review. Experimental Agriculture. 2011;**47**(4): 653-676. DOI: 10.1017/S0014479711000421

[13] Penman HL. Natural evaporation from open water, bare soil and grass. Proceedings of the Royal Society of London. 1948;**193**:120-145. DOI: 10.1098/rspa.1948.0037

[14] Allen RG, Pereira LS, Smith M, Raes D, Wright J. FAO-56 dual crop coefficient method for estimating

evaporation from soil and application extensions. Journal of Irrigation and Drainage Engineering. 2005;**131**:2-13

[15] Doorenbos J, Pruitt WD. Crop water requirements (revised). In: Food and Agriculture Organization of the United Nations, Irrigation and Drainage Paper 24. Rome: FAO; 1977

[16] Ventura F, Spano VF, Duce P, Snyder RL. An evaluation of common evapotranspiration equations. Irrigation Science. 1999;**18**:163-170

[17] Allen RG, Pereira LS, Raes D, Smith M. Crop evapotranspiration: Guidelines for computing crop water requirements. In: FAO Irrigation and Drainage Paper 56. Rome, Italy: FAO—Food and Agriculture Organization of the United Nations; 1998. 300 p

[18] Moser A, Leuschner C, Hartel D, et al. Response of cocoa trees (T cacao) to a 13-month dessication period in Sulawesi, Indonesia. Agroforestry Systems. 2010;**79**:171-187

[19] Diczbalis Y, Lemin C, Richards N, Wicks C. Producing cocoa in northern Australia. In: Australian Government, Rural Industries Research and Development Corporation Report 09/092. 2010

[20] Agele S, Famuwagun B, Ogunleye A. Effects of shade on microclimate, canopy characteristics and light integrals in dry season field-grown cocoa (*Theobroma cacao* L.) seedlings. Journal of. Horticulture. 2016;**11**(1):47-56

[21] Doorenbos J, Pruitt WD. Crop water requirements (revised). In: Food and Agriculture Organization of the United Nations, Irrigation and Drainage Paper 24, Rome. 1977

[22] Taku M, Saitoh TM, Nagai S, Noda HM, Muraoka H, Nasahara KN. Examination of the extinction coefficient in the beer–Lambert law for an accurate estimation of the forest canopy leaf area index forest. Science and Technology. 2012;**8**(2):67-76

[23] Sezen SM, Celikel G, Yazar A, Tekin S, Kapur B. Effect ofirrigation management on yield and quality of tomatoes grown indifferent soilless media in a glasshouse. Scientific Research and Essay. 2010;**5**:41-48

[24] Daymond AJ, Hadley P, Machado RCR. Canopy architecture, photosynthesis and yield of cocoa trees. Cafe, Cacao, The. 1992;**36**(2):103-108

[25] de Alvim PT. Cacao. In: Kozlowski TT, editor. Ecophysiology of Tropical Crops. 1977. pp. 279-313

[26] Goudriaan J, Monteith JL. A mathematical function for crop growth based on light interception and leaf area expansion. Annals of Botany. 1990;**66**: 695-701

[27] Acheampong K, Hadley P, Daymond AJ. Photosynthetic activity and early growth of four cacao genotypes as influenced by different shade regimes under west African dry and wet season conditions. Experimental Agriculture. 2013;**49**:31-42

[28] Kohlerlscher M, Schwendenmann L, Holscher D. Throughfall reduction in a cacao agroforest: Tree water use and soil water budgeting. Agric & Forest Meteorol. 2010;**150**:1079-1089

[29] Agele SO, Iremiren GO, Ojeniyi SO. Evapotranspiration, water use efficiency and yield of rainfed and irrigated tomato in the dry season in a humid rainforest zone of Nigeria. International Journal of

Biology & Agricultural Sciences. 2011;**13**: 469-476

[30] Famuwagun B, Agele S, Aiyelari P. Shade effects on growth and development of cacao following 2 years of continuous dry season irrigation. International Journal of Fruit Science. 2017;**18**(7):1-24

[31] Yang Q, Wang Y, Jia X, Zheng Y, He S, Deng L, et al. Fruit yield and quality response of Newhall navel orange to different irrigation regimes and ground cover in Chongqing three gorges reservoir area. Scientia Horticulturae. 2018;**241**(15):57-64. DOI: 10.1016/j.scienta.2018.06.083

[32] Orgaz F, Testi L, Francisco J, Villalobos LJ, Fereres E. Water requirements of olive orchards-II: Determination of crop coefficients for irrigation scheduling. Irrigation Science. 2006;**24**(2):77-84. DOI: 10.1007/s00271005-0012-x

[33] Bonachela S, Villalobos SJ, Fereres E. Soil evaporation from drip-irrigated olive orchards. Irrigation Science. 2001; **20**:65-71. DOI: 10.1007/s002710000030

[34] United Nations Framework Convention on Climate Change (UNFCCC). UNFCCC process: Convention, Kyoto Protocol and the Paris Agreement. 1994. Available from: https://unfccc.int

Chapter 7

Revolutionizing Rice Farming: Maximizing Yield with Minimal Water to Sustain the Hungry Planet

Shanmugam Vijayakumar, Narayanaswamy Nithya, Pasoubady Saravanane, Arulanandam Mariadoss and Elangovan Subramanian

Abstract

Increasing rice yield while reducing water usage is crucial to feed growing population. This chapter explores techniques to enhance irrigation efficiency and water productivity in rice farming while minimizing negative impacts like groundwater depletion, land subsidence, saltwater intrusion, and soil degradation. Modern techniques for rice farming bring significant benefits by increasing productivity, reducing water usage, and conserving natural resources. Promising techniques include direct-seeded rice, aerobic rice, drip-irrigated rice, saturated soil culture, IoT-based automated irrigation, and the system of rice intensification (SRI). For example, drip-irrigated rice increases yield by up to 20% using 30–50% less water, and the SRI boosts yield by up to 50% with 25–50% less water. Implementing these techniques improves rice productivity, income, food security, and water conservation. However, effectiveness varies based on soil, climate, labor force, and socio-economic status. Selecting suitable water-saving methods is crucial for maximizing farmer livelihoods while ensuring environmental safety.

Keywords: direct seeded rice, drip irrigated rice, automated irrigation, alternate wetting and drying, saturated soil culture

1. Introduction

India, with limited freshwater resources (only 4% of world's fresh water) is faced with the challenge of supporting its 1.39 billion population (20% of the global population) and their water demands. The three main water-consuming sectors in 1977 were agriculture (376.1 Billion Cubic Meters (BCM)), industry (14.16 BCM), and domestic use (12.46 BCM), totaling 403.3 BCM. Over the years, this has steadily risen to around 761 BCM in 2018, with agriculture alone making up 90.4% of the water consumption [1]. India's per capita water availability has decreased dramatically from 5600 m^3 in 1951 to 1481 m^3 in 2021 and is projected to decline further to below the water-stressed norm of 1700 m^3 per year by 2051. This decrease in water availability is rapidly transitioning India from a water-stressed to a water-scarce country, with many regions facing compromised water supplies to its citizen. The majority of India's water stress is due to poor water management rather than physical scarcity [1]. According to National Institution for Transforming India (NITI) Aayog (2018), the country faces a significant water crisis, with most cities lacking a 24-hour supply of drinking water and most households lacking access to safe drinking water. Approximately, 75% of residences lack on-premises access to safe drinking water, while 84% of rural households lack access to piped water. The growing population, rapid urbanization, and industrialization, along with the increased cropping intensity, has resulted in competing demand for water in India [2]. There is a growing concern among the citizens of India regarding an impending water crisis in the country.

India has 53.4% of its total land area under arid (15.8%) and semi-arid (37.6%) regions, where water scarcity deters agricultural growth and development. More than 45% of annual agricultural production comes from 18 to 20% of the world's irrigated land using 70% of all freshwater withdrawals. Irrigated agriculture is at least twice as productive per unit of land as compared to rainfed agriculture, enabling for more crop diversification and intensification in irrigated areas. Rice is the staple food for more than half of the world's population, and as such, its production has a significant role in achieving global food security. Rice is one of the most water-intensive crops and it is estimated that rice cultivation accounts for up to 30% of the world's freshwater withdrawals, with some regions heavily reliant on it for their water supply in [3]. However, traditional rice farming practices have resulted in inefficient use of resources, environmental degradation, and lower yields. Some of the major problems of traditional rice farming practices include *Overuse of water*: Traditional rice farming practices involve the flooding of rice paddies with water to control weeds and pests. However, this method leads to the overuse of water and environmental problems such as water pollution and degradation of aquatic habitats [4]. Low *yields*: Traditional rice farming practices often result in low yields due to inefficient use of fertilizers and other inputs. *Soil degradation*: Traditional rice farming practices can lead to soil degradation due to puddling which facilitate hardpan development, and continuous use of chemical fertilizers and pesticides, which harm beneficial organisms and promote soil degradation [5]. *Dependence on external inputs*: It often relies on application of external inputs such as chemical fertilizers, pesticides, and herbicides, which can be expensive and lead to environmental pollution [6]. *Labor-intensive*: Traditional rice farming practices are often labor-intensive, requiring significant manual labor for completing tasks such as planting, weeding, and harvesting. This can be a significant constraint for small-scale farmers with limited access to labor. *Vulnerability to climate change*: Traditional rice farming practices are often vulnerable to the impacts of climate change, such as droughts, floods, and pest outbreaks [7]. *Economic inefficiency*: Traditional rice farming practices can be economically inefficient due to the high cost of inputs, low yields, and dependence on external inputs [8].

1.1 Declining groundwater table

The extraction of groundwater increased as farmers shifted from the use of monoblock water pumps in the 1980s to submersible pumps to irrigate paddy crop. In addition, the incentivizing of irrigation leads to irrational increase in the number of electric tube wells and subsequently consistent drop in groundwater table. In the last two decades, the groundwater level dropped drastically to 150 ft. to 200 ft. from 30 ft. to 60 ft. in most parts of Punjab due to indiscriminate extraction [9].

Over exploitation has made groundwater not only scarce but also increasingly alkaline. Farmers often borrow money from informal and formal sources to install tube wells and purchase powerful submersible pumps [10]. This has caused indebtedness resulting farmers suicides in many places.

1.2 Food security nexus water availability

According to the NITI Aayog's Composite Water Management Index (CWMI) report-2019, India is facing a significant risk to its food security. Also, the country's top ten agricultural producing states are struggling to manage their water resources effectively. India has been classified as the highest drought-prone country in the world by the world water resources institute in 2019 based on drought intensity, water stress, and vulnerability to drought, population, crop, and livestock density. Out of 634 districts in India, only 38% (241 districts) are resilient to drought, while the rest are vulnerable to varying degrees of severity: mild (180 districts), moderate (80 districts), and severe (40 districts). Only ten states in the country have more than 50% of areas that are resilient to drought or dry conditions. The water quality of rivers and lakes in India is poor, with most of the river water (70%) being unsafe for consumption due to contamination from various sources, including industries, urbanization, and agricultural chemicals [11]. As a result, India ranks 120th out of 122 countries in the Water Quality Index [1].

1.3 Climate change nexus agriculture and its impact on water resource

Climate change is affecting rainfall patterns, and droughts are becoming more frequent and severe in many rice-growing regions. By increasing irrigation use efficiency, farmers can become less reliant on rainfall and more resilient to the effects of climate change. It also reduces the amount of water, energy, and other inputs required for rice cultivation, resulting in a more sustainable production system. Droughts are becoming more severe and persistent, exacerbated by climate change, in both irrigated and rainfed agriculture around the world, posing a greater risk to rural livelihoods by reducing crop and livestock output [7, 12]. This emphasizes the need for water resource development, conservation, and efficient use.

1.4 Low water productivity

India has one of the lowest values of water use efficiency (US$ 1.9/m^3) as compared to countries like Australia, USA, Brazil, Malaysia, China, South Africa, Mexico, Turkey and many Southeast Asian countries, which have an average water use efficiency of $15 per cubic meter [13]. Rice, a crop that requires a large amount of water, accounts for 50% of freshwater usage in agriculture. The water use efficiency of rice is also very low and the lowest among field crops [14]. In the state of Punjab, despite facing severe water stress, rice production relies entirely on irrigation and has a concerning annual groundwater extraction rate of 165.8%. Similarly, the trend of groundwater depletion is severe in Haryana (136.9%), Rajasthan (139.9%) followed by Tamil Nadu (80.9%) [15].

1.5 Deteriorating water quality

Pollution caused by various factors such as urbanization, industrialization, and the increased use of fertilizers and pesticides has led to significant deterioration in both surface water and groundwater quality in India. A major cause of concern is the discharge of untreated or inadequately treated industrial waste, effluents, and sewage into rivers and rivulets, which demands immediate attention [16]. Unfortunately, traditional water bodies like ponds and wells in rural areas are also at risk. Many have already been filled or encroached upon, while others have been turned into dumping grounds [17]. To make matters worse, the mixing of stormwater and sewage in many municipal towns has resulted in the contamination of water resources. A report suggests that nearly 50–60% of groundwater is still fresh and fit for consumption, 20–30% is moderately saline and of marginal quality. However, 15–25% of groundwater is saline, alkaline, and unsuitable for irrigation.

As water resources become increasingly scarce, finding sustainable solutions to manage and allocate this precious resource is becoming more critical. Moreover, irrigation accounts for a significant proportion of rice production cost, making it essential to improve irrigation efficiency to address the water crisis while simultaneously increasing yields and reducing production costs. Several agronomic technologies and production methods are available for rice that can enhance water productivity. However, each technology comes with its own practical difficulties, despite its ability to increase grain yield, save water, and reduce GHG emissions. Therefore, it is essential to study the advantages, disadvantages, and future research needs of each technology to make informed decisions about their use.

2. Technologies for saving irrigation water in rice cultivation

2.1 System of Rice Intensification

In recent years, the System of Rice Intensification (SRI) has gained attention as a revolutionary approach to rice production, offering significant opportunities for farmers to increase yields, reduce input costs, and improve environmental sustainability [18]. The SRI is a holistic approach to rice cultivation that emphasizes principles such as transplanting single, young seedlings (13–15 days old), keeping the soil moist rather than flooded, cono weeding, using organic inputs, and using wider spacing (25 × 25 cm) between plants (**Figure 1**). The approach is founded on the principle that rice plants do not need to be flooded to grow optimally, and by managing the plant's growth environment more effectively, farmers can achieve higher yields with fewer inputs. These practices promote vigorous root growth, which in turn leads to healthier plants and higher yields [19]. The method also promotes the use of locally available inputs and promotes organic farming practices. By using fewer inputs and promoting soil health, SRI can also reduce input costs, making it a more economically viable approach for small-scale farmers [6]. SRI has proven to be a highly effective approach that offers significant environmental and economic advantages. With over a 60% increase in yield, it has also helped to reduce GHG emission by 40%, ground-water usage by 60%, and fossil energy consumption by 74% per kilogram of rice produced [20]. A review of 78 studies comparing SRI to traditional rice cultivation methods found that SRI resulted in higher grain yields in 80% of evaluations, yielding 24% more than Best Management Practices (BMPs) and 56% more than Farmer Practices (FPs) while reducing input of seed, water, and fertilizer, making it an effective method for increasing production with fewer resources [21]. The evaluation of SRI in India found that it increased yield by 60%, while reducing net GHG emissions by 40%, decreasing groundwater depletion by 60%, and reducing fossil-energy use by 74% [20].

Figure 1.
System of rice intensification.

However, despite its potential benefits, the adoption of SRI has been limited due to several constraints. One of the most significant constraints is the lack of awareness and understanding of the SRI approach among farmers and extension workers. This has led to limited adoption and scaling of the approach, particularly in areas where traditional rice farming practices are deeply entrenched. Another constraint is the lack of access to appropriate inputs. SRI requires inputs such as compost and organic matter, which may not be readily available or affordable for many small-scale farmers [6]. Additionally, the approach requires significant changes in traditional rice farming practices, which may be difficult for some farmers to adopt. Despite these challenges, there are several actions that can be taken to promote the adoption and scaling of the SRI approach. First, there is a need for increased awareness and education about the approach among farmers and extension workers. This can be achieved through training and education programs that highlight the benefits of the approach and provide guidance on how to implement it effectively [22]. Secondly, there is a need for increased investment in research and development to identify and promote appropriate inputs and technologies for SRI. This can include the development of locally appropriate composting techniques and the use of innovative technologies such as precision agriculture [22]. Thirdly, there is a need for policy and institutional support to promote the adoption and scaling of SRI. This can include the provision of financial incentives and technical support for farmers to adopt the approach, as well as the establishment of policies and regulations that can promote sustainable and environmentally friendly farming practices.

2.2 Saturated soil culture

Rice cultivation has traditionally relied on flooding the fields with water to create an anaerobic environment for the rice plants. However, this method is not sustainable and can lead to significant water loss through percolation. To address this issue, alternative methods viz. saturated soil culture (SSC) has been developed. In SSC, the water loss through percolation is significantly lower because the soil is kept at saturation level, without flooding the field. SSC also reduces the need for frequent irrigation (**Figure 2**). This makes SSC a more sustainable and water-efficient method of rice cultivation compared to conventional methods. In addition, SSC has several advantages over conventional flooded rice cultivation, including higher yields, reduced greenhouse gas emissions, improved nutrient use efficiency, and improved soil health. In SSC the crop roots are supplied with sufficient oxygen, which can improve crop growth and productivity. SSC has been shown to increase water use efficiency by reducing the loss of unproductive water outflows, allowing more water to be utilized by the rice plants [23]. It is important to note that the yield benefits of SSC can be impacted by various factors, such as water management practices, soil type, and climatic conditions. Careful management of water levels, nutrient inputs, and pest and disease control is necessary to maximize the yield potential of SSC. SSC can increase rice yields by up to 50% compared to traditional upland rice cultivation methods [24]. Sometimes, implementing the SSC practice may cause a decline in rice yields if irrigation is not managed properly [25]. Since SSC involves the alternate wetting and drying of the soil, weed growth can be a problem. Farmers may need to implement additional weed management practices to control weed growth [26]. In areas where the water used for irrigation is saline, SSC can lead to the accumulation of salt in the soil, which can reduce crop growth and productivity. SSC is suitable for regions with high rainfall or access to irrigation water, as the soil needs to be kept consistently wet. It is also best suited to soils that have good water-holding capacity, such as clay soils. Since the water level is maintained just below the soil surface, the methane emission from SSC is lower than those from traditional flooded rice cultivation. The impact of SSC on GHG emissions and yield depends on various factors, including the management practices used and the soil type [4]. However, SSC can lead to increased emissions of nitrous oxide, a potent GHG, due to the intermittent flooding and alternate wetting and drying.

Compared to traditional farmer practice, the use of saturation soil irrigation resulted in a reduction of 10, 18, and 14% in plant height, tillers, and leaves number,

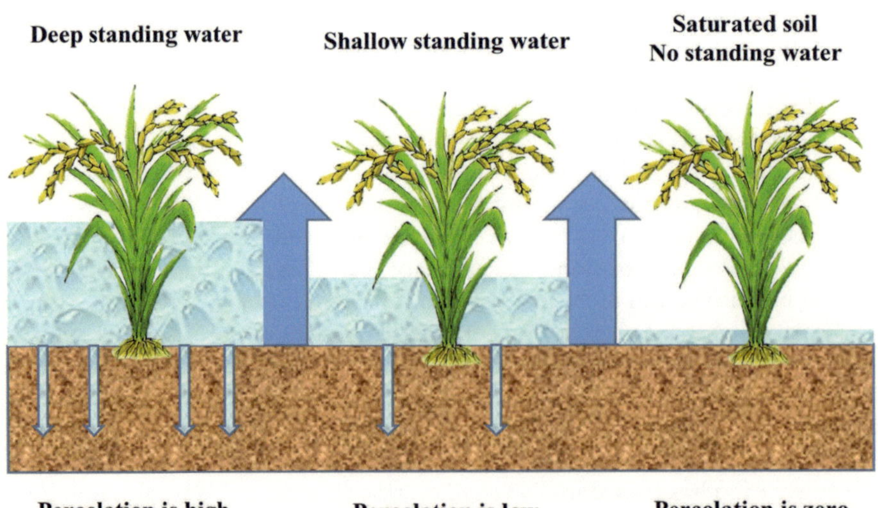

Figure 2.
Saturated soil culture.

respectively. However, when soil was saturated weekly at 120%, it showed higher irrigation productivity (0.69 kg/m^3), rainwater productivity (1.02 kg/m^3), and water-saving (90.53%) with minimal yield reduction (5×10^{-3} kg/m^3) [27]. Yanti et al. conducted a study to investigate the impact of water availability on the growth of local rice varieties. The results indicated that plant height was unaffected by water availability, while the number of tillers was influenced. Tiller formation was observed during the vegetative phase, approximately 28 days after planting, until the start of the generative phase, around 55 days after planting. During the generative phase, unproductive tillers were eliminated or dried up [28].

2.3 Aerobic rice system

Aerobic rice approach involves cultivating rice in aerobic soil conditions, with lower water inputs and without flooding the fields. Aerobic rice cultivation requires 30–50% less water than traditional flooded rice paddies, making it a more sustainable and environmentally friendly approach [29]. By eliminating flooding, aerobic rice cultivation significantly reduces methane emissions, a potent greenhouse gas produced by traditional rice farming (**Figure 3**). Although aerobic rice systems have the potential to produce yields similar to or higher than those of traditional flooded rice systems, the success of the system may be influenced by a range of factors, including climate, soil conditions, and crop management practices. Proper soil management is critical to successful aerobic rice cultivation, including effective weed and pest control and nutrient management [30]. The adoption of aerobic rice systems requires changes in crop management practices, such as nutrient management, weed management, and planting density. Farmers may need training and support to adopt these new practices successfully. Aerobic rice systems require well-drained soils with good organic matter content. In areas where soil is poorly drained or low in organic matter, additional soil management practices may be

Figure 3.
Aerobic rice system of rice cultivation.

necessary to ensure the success of aerobic rice systems. In an aerobic rice system, the presence of oxygen in the soil promotes the growth of aerobic microorganisms that decompose organic matter at a faster rate compared to anaerobic decomposition in waterlogged conditions [31]. This faster rate of decomposition can result in a decrease in soil organic carbon content over time, as compared to soil under waterlogged conditions. However, an aerobic rice system may offer other benefits such as improved plant productivity, better nutrient use efficiency, and reduced greenhouse gas emissions. Soil submergence or waterlogging in rice leads to less oxygen, favoring slow anaerobic decomposition and buildup of organic carbon, improving soil fertility. It also accumulates sediment and nutrients, altering soil structure and nutrient availability, affecting plant growth and ecosystem health. However, prolonged submergence may reduce productivity and increase disease susceptibility [32]. According to the survey conducted among aerobic rice farmers, several factors favored the adoption of the system, such as ease of operation due to direct seeding, an increase in resource-use efficiency, particularly for labor, higher net profitability, water-saving through reduced irrigation, and the option for crop diversification in mixed-cropping systems. Conversely, factors that discouraged repeat plantings included the unavailability of suitable fine grain basmati varieties, problems with weeds and diseases, poor germination, spikelet sterility, frequent irrigation requirements, low yields, and unsuitable soil types [33].

Here are some potential areas for new research in aerobic rice systems: *Crop management practices*: There is a need to develop crop management practices that are specific to aerobic rice cultivation, such as optimizing fertilizer use, irrigation scheduling, weed control, and pest management. In addition, research is needed to identify the most suitable rice varieties for aerobic conditions, and to develop improved breeding strategies for these varieties. *Soil health*: The long-term effects of aerobic rice cultivation on soil health and fertility are not yet fully understood, and there is a need for research to evaluate the impacts of different management practices on soil properties, including nutrient availability, organic matter content, and soil structure [31, 32]. *Climate resilience*: As aerobic rice systems are more susceptible to drought and other climate-related stresses than flooded systems, research is needed to identify strategies to increase the resilience of these systems to climate change, such as the development of drought-tolerant rice varieties or the use of conservation agriculture practices. *Water use efficiency*: Aerobic rice systems typically use less water than flooded systems, but there is a need for research to optimize irrigation practices and improve water use efficiency in these systems. *Economic viability*: The economic viability of aerobic rice system compared to flooded rice systems is not yet fully understood. Research is needed to evaluate the economic costs and benefits of aerobic rice cultivation, and to identify strategies to improve profitability and marketability of the produce.

2.4 Alternate wetting and drying

Alternate wetting and drying (AWD) technique involves allowing the paddy fields to dry out periodically instead of keeping them continuously flooded. This method is gaining popularity among rice farmers because it can reduce water use while maintaining or even increasing crop yields (20%), reducing methane emissions by up to 50% and increase oxygen availability to the roots [34]. In AWD, farmers must monitor the soil moisture level using a water level indicator (**Figure 4**). Once the water level reaches a certain point, the farmer stops the irrigation and allows the soil

Figure 4.
Alternate wetting and drying.

to dry. When the soil reaches a certain dryness level, the farmer resumes irrigation. This process is repeated throughout the growing season, depending on the rice variety and weather conditions. AWD requires careful monitoring of the soil moisture level to avoid under or over-irrigation. IRRI's "safe AWD" recommendations, intended to minimize yield reductions, are that the soil is dried until soil water depth reaches 15 cm below the surface and that the field is re-irrigated to a standing water depth of around 5 cm [35].

The benefits of AWD can vary depending on the rice variety, soil type, and weather conditions. Therefore, farmers may need to adjust their irrigation practices accordingly.AWD can increase the risk of yield loss if the soil is allowed to dry out too much, particularly during critical stages of crop development such as flowering and grain filling. If the soil is too dry, it can result in reduced nutrient uptake, decreased photosynthesis, and decreased grain yield. AWD can also lead to increased weed growth if the field is drained for too long. Here are some potential areas for new research in AWD: *Water management*: There is a need to develop optimized water management practices for AWD, including determining the optimal timing and duration of the wet and dry periods, and the most suitable irrigation methods to use. In addition, research is needed to identify the most effective water-saving strategies for AWD, such as the use of mulching, cover crops, or different soil types. *GHG emissions*: AWD has been shown to reduce methane emissions from rice cultivation. However, there is a need for further research to understand the mechanisms behind this reduction and to quantify the overall GHG emissions impact of AWD compared to traditional flooded rice cultivation. *Economic viability*: Research is needed to assess the economic costs and benefits of AWD and to identify strategies to improve its profitability and marketability.

2.5 Drip irrigated rice

Drip irrigation is well-suited for rice cultivation as it allows precise control of water application, which is essential for this water-intensive crop. Irrigation water is supplied to the crop through drip emitters that are placed at regular intervals along the field (**Figure 5**). The spacing between emitters depends on the soil type, crop variety, and other factors, and can range from 30 cm to 60 cm [36]. The water is delivered slowly and evenly to the root zone of the plants, which reduces water loss through evaporation and runoff. It facilitates the precise application of fertilizers and other inputs, which can further improve crop performance. However, drip irrigation

Figure 5.
Drip irrigated rice.

requires careful management and maintenance to ensure that the system is functioning properly and that the plants are receiving the appropriate amount of water [37]. Factors such as soil moisture, crop growth stage, and weather conditions must be monitored regularly to adjust the irrigation schedule as needed. The timing and amount of irrigation in drip irrigated rice depend on several factors, including soil type, climate, crop stage, and water availability. The frequency of irrigation in drip irrigated rice depends on the rate of water uptake by the plants, which is influenced by factors such as temperature, humidity, wind, and plant density. In general, drip irrigated rice should be irrigated when the soil moisture content in the root zone reaches the lower limit of the desired range. This can be determined by monitoring soil moisture using a soil moisture sensor or by checking the soil moisture by hand. Similarly, the amount of water applied should be sufficient to replenish the soil moisture deficit and bring the soil moisture content to the upper limit of the desired range. The desired range depends on the soil type and can range from 60 to 80% of field capacity [38].

Drip irrigation reduces weed growth compared to flood irrigation, but it can also result in more weeds growing directly under the drip lines. The use of drip irrigation resulted in a 29% increase in aerobic rice yield and a 50% reduction in water usage and the water productivity of aerobic rice under drip irrigation showed a twofold increase compared to other irrigation methods. Drip irrigation for direct-seeded rice increased yield (7.34–8.01 t ha^{-1}) and water-use efficiency (0.81–0.88 kg m^{-3}) while reducing water use by over 40% compared to flood irrigation (6.63–7.60 t ha^{-1} and 0.42–0.52 kg m^{-3}, respectively). Root density at 15–30 cm soil depth was also higher in drip-irrigated crops (0.86–1.05 mg cm^{-3}) than in flood-irrigated crops (0.76–0.80 kg cm^{-3}) [39]. Among different drip irrigation configurations tested, the subsurface drip with a lateral distance of 0.8 m and 1.0 L/h dripper discharge irrigation system demonstrated the best performance in terms of rice growth, physiology, and yield [40].

Drip fertigation can be an effective and sustainable method for fertilizing crops, providing precise and uniform delivery of nutrients while reducing nutrient, labor and water use [41]. Subsurface drip irrigation in rice-wheat cropping system reduces the nitrogen requirement by 20% in both the crop and obtain grain yields similar to that of flood irrigated crops [42]. However, it requires careful planning and management to ensure proper functioning and effective nutrient application. Drip fertigation requires specialized equipment, such as a fertigation pump, injection system, and filters. This equipment must be properly installed and maintained to ensure proper functioning. The quality of the irrigation water can impact the effectiveness of drip fertigation. High levels of dissolved salts, bicarbonates, or other impurities can lead to clogging of the emitters or reduced fertilizer uptake by plants [43]. The timing and rate of fertilizer application through drip fertigation must be carefully adjusted based on the crop stage and nutrient requirements. Over-application of fertilizers can lead to nutrient imbalances, reduced crop quality, and environmental pollution.

2.5.1 Prospects of drip irrigated rice

Drip irrigation has several advantages over traditional flood irrigation methods for rice cultivation. *Water use efficiency:* Drip irrigation can reduce water use by up to 50% compared to traditional flooded rice fields. This can lead to cost savings for farmers and help conserve water resources. *Higher yields*: Drip irrigation can improve rice yields by ensuring a consistent and adequate supply of water and nutrients to the plants. This can lead to higher grain quality and quantity. *Better fertilizer management*: Drip irrigation can improve fertilizer use efficiency by delivering nutrients directly to the root zone of the plants. This can reduce fertilizer leaching and runoff, which can harm the environment. Drip irrigation combined with fertigation (the application of fertilizers through irrigation water) significantly improves rice yield and nutrient use efficiency. *Reduced labor requirements*: Drip irrigation requires less labor for irrigation and weeding than traditional flooded rice fields, which can lead to cost savings and increased efficiency. *Climate resilience*: Drip irrigation can help rice farmers adapt to climate change by providing more precise control over water and nutrient delivery. This can help reduce the risk of drought and flood damage to rice crops. *GHG emission*: Drip irrigation can significantly reduce methane emissions from rice fields, compared to flood irrigation. Based on two-year average, Fawibe et al. found that the global warming potential is reduced by 89% under drip irrigated plastic mulch system compared to conventional flooding [44]. *Soil compaction*: Drip irrigation in rice cultivation helps in improving soil health by reducing soil erosion and compaction.

2.5.2 Constraints of drip irrigated rice

High initial investment: Drip irrigation systems require significant initial investment in equipment and infrastructure, such as pumps, filters, and drip lines. This can be a significant barrier for small-scale rice farmers. *Technical knowledge and skills*: Drip irrigation requires technical knowledge and skills in installation, operation, and maintenance. The system must be regularly maintained to ensure proper functioning, which can involve tasks such as cleaning filters, replacing emitters, and checking for leaks. This is a time-consuming and labor-intensive process. Farmers may need training and support to adopt this method successfully. Soil *suitability*:

Drip irrigation is most suitable for well-drained soils with good water-holding capacity. Therefore, its use may be limited to regions with heavy clay soils or low water-holding capacity. *Pest and disease management*: Drip irrigation may increase the incidence of some pests and diseases, such as root rot and nematodes. Therefore, farmers need to implement appropriate pest and disease management practices to mitigate these risks. *Energy requirements*: Drip irrigation systems require energy for pumping and filtration, which can increase production costs and contribute to greenhouse gas emissions [45]. *Clogging*: Clogging of drip system due to the buildup of sediment, organic matter, or mineral deposits in the emitters or filters reduces the effectiveness of the system, leading to uneven application of fertilizers and water [43]. *Water quality*: Poor quality irrigation water requires additional treatment or management.

2.5.3 Potential areas for new research in drip irrigated rice

Drip irrigation has the potential to improve water use efficiency, increase rice yields, and reduce labor requirements. However, its adoption may depend on several factors, such as the availability of appropriate soil and water resources, farmer knowledge and skills, and the initial investment cost. Therefore, further research and extension efforts are needed to promote the adoption of drip irrigated rice among rice farmers worldwide. Here are some potential areas for new research in drip irrigated rice. *Water and nutrient management*: Drip irrigation provides precise control over water and nutrient application, but research is needed to develop optimized water and nutrient management practices for drip-irrigated rice, including determining the most suitable timing and quantity of water and nutrient application to maximize yield and minimize water and nutrient losses [36]. *Crop management*: Drip irrigation can affect crop management practices such as weed control, pest management, and disease control, and research is needed to develop integrated crop management practices for drip-irrigated rice that can optimize yield and reduce production costs. *Soil health*: Drip irrigation can impact soil health and nutrient availability, and research is needed to evaluate the long-term effects of this technique on soil properties, including organic matter content, nutrient availability, and soil structure. *Energy use and greenhouse gas emissions*: Drip irrigation requires energy to operate, and the energy use and associated greenhouse gas emissions of drip-irrigated rice need to be quantified and compared to traditional flooded rice cultivation. *Economic viability*: Drip irrigation has the potential to increase water and nutrient use efficiency and yield stability, but its economic viability compared to traditional flooded rice cultivation needs to be further evaluated. Research is needed to assess the economic costs and benefits of drip-irrigated rice and to identify strategies to improve its profitability and marketability.

2.6 Automated irrigation

Automated irrigation using sensors and the Internet of Things (IoT) can be an effective way to save water in rice cultivation as it requires large amounts of water to grow. This approach involves using sensors to collect data on various environmental factors such as temperature, humidity, soil moisture, and weather conditions. The data is then transmitted to a central hub via the internet, which processes the information and triggers irrigation systems as needed (**Figure 6**).

Figure 6.
Automated smart irrigation system.

In traditional irrigation methods, there is a risk of overwatering or underwatering, which can lead to reduced crop yield, increased water usage, and higher costs [46]. By using sensors and IoT, farmers can optimize water usage and reduce waste by ensuring that the rice fields receive only the necessary amount of water. Automated irrigation systems can also help farmers save time and labor by eliminating the need for manual monitoring and adjustment of irrigation systems [47]. This allows farmers to focus on other important aspects of rice cultivation such as soil management, pest control, and harvesting. The implementation of automatic irrigation system increased on-site water productivity by 12.7%, and the labor power required for water management decreased by 21.8%. Moreover, the investment in the automatic irrigation system yielded positive financial returns, with an internal rate of return of 8.6% higher than the discount rate of 4.5% and benefit-cost ratio was 1.23 [48].

The modern drip irrigation system is highly efficient in water usage when compared to traditional irrigation methods. Automated drip irrigation can save up to 41.5% of water compared to conventional flood irrigation and 13% compared to traditional drip irrigation methods [49]. Sensors used in automated irrigation help in accurate weather predictions which assist farmers in designing optimal irrigation schedules that consider expected rainfall and evapotranspiration rates. The high accuracy weather prediction can help in preventing overwatering, which can waste water and reduce water use efficiency [50]. In addition, farmers can take preventive measures such as adjusting irrigation schedules and creating adequate drainage facilities in response to extreme weather events like droughts, floods, and storms. Moreover, better weather prediction can aid farmers in planning the timing of fertilizer and pesticide applications. This can enhance nutrient uptake and pest management, resulting in improved crop growth and water use efficiency.

2.6.1 Bottleneck for automated irrigation

There are several bottlenecks that need to be considered when implementing automated irrigation systems in rice cultivation: *Cost*: Implementing an automated irrigation system can be expensive, especially for small-scale rice farmers. The cost of the sensors, communication systems, and other equipment needed to set up the system can be a barrier to adoption [51]. *Technical expertise*: Setting up and maintaining an automated irrigation system requires technical expertise. Farmers may need training to operate the system and troubleshoot any technical issues that may arise. *Infrastructure*: The implementation of automated irrigation systems requires reliable internet connectivity and power supply, which may not be available in some rural areas where rice cultivation is prevalent [52]. *Compatibility*: Not all automated irrigation systems may be compatible with the specific needs of rice cultivation. Rice cultivation requires different water management practices at different growth stages, which may need to be factored in when setting up the system. *Data analysis and interpretation*: Collecting data through sensors is not enough; the data also needs to be analyzed and interpreted to trigger irrigation events effectively. Farmers may need assistance with data analysis and interpretation to ensure that the system is functioning optimally [53]. *Cultural factors*: Some farmers may be resistant to adopting new technology or changing traditional irrigation practices, which could limit the adoption of automated irrigation systems.

Here are some potential areas for future research in automated irrigation for rice: *Sensor technology*: The effectiveness of automated irrigation systems depends on the accuracy of the sensor technology used to measure soil moisture and other environmental variables. Future research is needed to develop and improve sensor technology for automated irrigation in rice, including sensors that can measure soil moisture, temperature, and other factors at various depths and locations within the field [54]. *Decision support systems*: Automated irrigation systems rely on decision support systems that use real-time data from sensors to make decisions about irrigation timing and quantity. Future research is needed to develop and improve decision support systems for automated irrigation in rice, including machine learning algorithms that can predict crop water demand and optimize irrigation management [55]. *Crop modeling*: Crop modeling can be used to simulate crop growth and water use under different irrigation scenarios and can be a valuable tool for optimizing automated irrigation systems. Future research is necessary to develop and improve crop models for rice cultivation that can accurately predict crop water demand and growth under different irrigation scenarios. *Energy efficiency*: Automated irrigation systems require energy to operate, and the energy use and associated greenhouse gas emissions of these systems need to be quantified and minimized. Future research is needed to optimize the energy efficiency of automated irrigation systems for rice cultivation, including the use of renewable energy sources such as solar power. *Economic viability*: The economic viability of automated irrigation systems for rice cultivation needs to be further evaluated, including the costs and benefits of different types of sensors, decision support systems, and energy sources [51]. Future research is needed to identify strategies to improve the profitability and marketability of automated irrigation systems for rice cultivation.

2.7 Direct seeded rice

Direct-seeded rice (DSR) is a method of rice cultivation where seeds are directly sown into the field instead of transplanting seedlings. This method is becoming increasingly popular as it saves labor, water, and time compared to traditional transplanting methods [56]. There are two types of DSR methods: wet DSR and dry DSR. In dry DSR, the rice seeds are broadcasted or drilled into dry soil (**Figure 7**), whereas in wet DSR, the pre-germinated rice seeds are sown on wet soil (**Figure 8**). Dry DSR resulted in higher yields (13–18%) and reduced total water inputs (8–12%) compared to TPR [57]. DSR has the potential to meet global demand while reducing water usage by 50%, labor costs by 60%, and increasing productivity by 5–10% [58]. Some of the common problems associated with DSR include poor seedling establishment, uneven crop emergence, weed competition, pest and disease infestations, and nutrient deficiencies. *Uneven crop emergence*: DSR can experience uneven crop emergence, leading to lower yields and reduced crop quality. This can be caused by factors such as uneven seed placement, variable soil moisture, and differences in soil temperature [59]. *Weed competition*: DSR is more vulnerable to weed competition than transplanted rice, as weeds can easily outcompete rice seedlings for nutrients and light [60]. *Pest and disease infestations*: DSR can be more susceptible to pest and disease infestations than transplanted rice, due to factors such as reduced plant vigor and higher weed pressure. *Nutrient deficiencies*: DSR can experience nutrient deficiencies, particularly during the early stages of growth, due to factors such as reduced nutrient availability and increased leaching [61]. *Water management:* DSR requires careful water management to ensure that the seeds do not dry out before germination, and that the young seedlings are not drowned by excessive flooding. *Labor requirements*: Direct-seeded rice can require more labor than traditional transplanting methods, particularly for weed control and crop establishment [62].

Figure 7.
Dry direct seeding of rice through seed drill.

Figure 8.
Wet direct seeding of rice using drum seeder.

2.7.1 Research needs for DSR include

Seed technology: Development of improved varieties that have better seedling vigor, higher germination rates, and tolerance to abiotic stresses such as drought and salinity [59]. *Weed management*: Development of effective and sustainable weed control strategies for DSR that minimize the need for herbicides and reduce the impact of weeds on crop yield [62]. *Nutrient management*: Optimization of nutrient management strategies for DSR that take into account the different nutrient requirements of direct-seeded rice and the impact of soil moisture on nutrient availability. *Pest and disease management*: Development of integrated pest and disease management strategies for DSR that reduce the incidence of pest and disease infestations and minimize the use of chemical pesticides. *Agronomic practices*: Development of agronomic practices that optimize crop establishment, promote uniform emergence, and reduce the risk of crop failure due to unfavorable weather conditions. *Economics and marketability*: Analysis of the economic viability of DSR compared to traditional transplanting methods and identification of strategies to improve the marketability of direct-seeded rice.

3. Other agronomic management practices

Farmers should adopt various strategies such as proper fertilization, timely and efficient weed management and the use of drought-tolerant crop varieties to improve crop growth, productivity, and water use efficiency. Building bunds or levees around the paddy fields can prevent water loss through runoff and seepage.

3.1 Best nutrient management

Balanced fertilizer application: Applying the right amount of fertilizers and balancing the nutrient supply with rice crop demand can improve the efficiency of water use [14]. Whereas, over-fertilization can result in excessive vegetative growth, leading to higher water demand, which ultimately results in low yield and reduced water use efficiency. Similarly, under-fertilization can negatively impact crop growth and yield potential, leading to lower water use efficiency. Therefore, maintaining an appropriate balance of fertilizers and nutrients is crucial for achieving optimal water use efficiency in rice cultivation [63].

Organic matter application: Incorporating organic matter, such as compost or manure, into the soil can improve soil structure and water holding capacity, which can increase water availability for plant uptake and reduce the need for irrigation.

Timing and placement of fertilizers: The timing and placement of fertilizers can affect the efficiency of water use. Applying fertilizers in split doses during the growing season, rather than in a single application, helps to reduce leaching losses and increase nutrient uptake efficiency [64]. Placing fertilizers near the root zone also improves nutrient uptake and reduces losses due to runoff or leaching. The increased uptake and efficient utilization of nutrients, in turn, promotes crop growth and yield, ultimately leading to improved water use efficiency.

Use of micronutrients: Micronutrients, such as zinc and iron, play a vital role in the growth and development of rice plants. Deficiencies of these micronutrients can reduce crop growth and yield potential, leading to lower water use efficiency. Supplementing the soil with these micronutrients can help to improve crop growth and yield potential, leading to more efficient use of water.

Nutrient deficiency is a common problem in improved rice production technologies like aerobic rice, direct seeded rice, drip irrigated rice and alternate wetting and drying [65]. Nutrient deficiency can limit the ability of crops to effectively utilize water resources. Nutrient deficiency occurs when plants do not receive adequate amounts of essential nutrients such as nitrogen, phosphorus, potassium, and micronutrients such as zinc, iron, and manganese. The lack of these essential nutrients can significantly impact crop growth and productivity, leading to reduced yields, poor quality produce, and increased susceptibility to pests and diseases which intern resulted in lower water use efficiency and water productivity. Thus, optimum fertilization based on soil fertility status and environmental factors could increase water use efficiency and water productivity.

3.2 Efficient weed management

Weeds compete with rice plants for water, nutrients, and sunlight. Thus, it is essential to manage weeds effectively to reduce their negative impact on crop growth, yield and water use efficiency [66]. When weed populations are high, they can reduce the amount of water available to the rice plants by increasing the water loss from the soil through transpiration. As a result, more water may need to be applied to the field to maintain the same level of soil moisture, leading to decreased water use efficiency. Weeds also interfere with the uptake of nutrients by rice plants. This can lead to reduced plant growth and development, which in turn can reduce water use efficiency. Weeds also act as hosts for many pests and diseases that can infect rice plants. This can lead to reduced plant growth and yield, and again, reduce water use efficiency. Controlling weeds can reduce their competition for water, nutrient, space and light, thereby increasing water availability, water use efficiency and yield of rice plants [67]. Therefore, it is more advantageous to implement early weed control measures such as pre-emergence herbicides, manual weeding, and mechanical weed removal, as opposed to late season weed control, to maximize water use efficiency in rice cultivation [68].

3.3 Selecting suitable cultivar

Planting drought-tolerant rice varieties can reduce the amount of water needed for cultivation while maintaining yields. Selecting drought-tolerant cultivars is an important strategy for improving water use efficiency in rice production [69]. Drought tolerant cultivars have better yield stability (able to maintain yield under water-limited conditions) than conventional cultivars. If a region is expected to experience a drought, farmers may choose to plant drought-tolerant rice varieties that can thrive in dry conditions. Drought-tolerant rice cultivars have several adaptations such as deep root system, smaller and thicker leaves, stomata closing during the hottest part of the day, osmoregulation, early maturity and improved photosynthesis that enable them to thrive well in dry conditions. By using less water, drought-tolerant cultivars can help in reducing the environmental impact of rice production. This includes reducing water consumption, greenhouse gas emissions, and nutrient runoff.

Globally several drought tolerant cultivars are developed in different countries (**Table 1**). Example, *BRS Primavera* cultivar was developed by the Brazilian Agricultural Research Corporation (EMBRAPA) is known for its high yield potential, tolerance to low soil fertility and shown to perform well in aerobic rice cultivation systems in Brazil. *NSIC Rc222* cultivar was developed by the Philippine Rice Research Institute (PhilRice) is known for its tolerance to drought, heat, and low

Cultivar	Duration	Yield (t/ha)	Ecology	Tolerance
Sahbhagi Dhan (DRR Dhan 42)	120 days	5.0–5.5	Upland and drought prone shallow lowlands	Drought tolerant
CR Dhan 801	140 days	Normal condition—6.3 Under submergence—4 Under drought—2.9	Climate-smart rice variety suitable for all ecology	Submergence as well as drought tolerance
CR Dhan 802	139 days	Normal condition—6.5 Under submergence—4.3 Under drought—2.3		
MAS 946-1 (Sharada)—Aerobic Rice Variety	120 days	5.51	Irrigated rice ecosystem	Drought tolerance
Rajalaxmi (CRHR-5)	130 days	6–7	Irrigated rice ecosystem	Submergence, cold at seedling stage, salinity/alkalinity tolerance

Table 1.
Drought tolerance rice varieties.

soil fertility. *Aerobic Rice 1* cultivar was developed by the International Rice Research Institute (IRRI) and is known for its tolerance to low soil fertility and high yields in non-flooded conditions. It has been tested and shown promising results in several countries in Asia. *IR72* cultivar was also developed by the IRRI and is known for its tolerance to drought and low soil fertility. *CT9993-5-10* cultivar was developed by the Chinese Academy of Agricultural Sciences and is known for its tolerance to drought and low soil fertility. WAB56-104: This cultivar was developed by the West Africa Rice Development Association (WARDA) and is known for its tolerance to drought, pests, and diseases. It has been shown to perform well in aerobic rice cultivation systems in West Africa. *Ariete* cultivar was developed by the International Center for Tropical Agriculture (CIAT) and is known for its tolerance to low soil fertility and high yields in non-flooded conditions. It has been tested and shown promising results in several countries in Latin America.

4. Conclusion

Revolutionizing rice farming to maximize yield with minimal usage of water is crucial for sustaining a hungry planet. The overuse of water for rice cultivation in conventional rice cultivation method, particularly in regions with limited water resources, has placed a significant strain on water availability and led to various environmental issues, including greenhouse gas emissions. The development of innovative farming techniques such as the SRI, DSR, AWD, SSC and aerobic rice system and the use of drought-resistant varieties of rice has shown promise in reducing water wastage while increasing yield. Smart irrigation utilizing sensors and IoT is poised to revolutionize irrigation management in rice cultivation. Furthermore, incentivizing the adoption of improved irrigation methods through carbon and water credits can promote water conservation in rice farming. In addition to water conservation, researchers are taking a systems approach to address various issues in rice cultivation, including labor, greenhouse gas emissions, weed control, nutrient use efficiency, pest and disease management, production cost, and cropping intensity. Many of these improved irrigation methods are addressing most of the problems in rice cultivation beside saving irrigation water and improving water productivity. It is essential to continue research and development efforts in this area to ensure food security for the growing population while preserving our precious water resources. Through collective efforts, we can work towards a sustainable future where we produce more with less water, ensuring food security for all.

Acknowledgements

Authors would like to extend sincere gratitude to the authors whose research findings have greatly contributed to the writing of this chapter.

Conflict of interest

All the authors declare that there are no potential conflicts of interest associated with this submission.

Author details

Shanmugam Vijayakumar[1*], Narayanaswamy Nithya[2], Pasoubady Saravanane[3], Arulanandam Mariadoss[4] and Elangovan Subramanian[5]

1 ICAR-Indian Institute of Rice Research, Hyderabad, India

2 Department of Seed Science and Technology, SRS Institute of Agriculture and Technology, Vedasandur, Tamil Nadu, India

3 Pandit Jawaharlal Nehru College of Agriculture and Research Institute, Karaikal, India

4 National Institute of Plant Health Management, Hyderabad, India

5 Krishi Vigyan Kendras, Tamil Nadu Agricultural University, Madurai, Tamil Nadu, India

*Address all correspondence to: vijitnau@gmail.com; vijayakumar.s@icar.gov.in

References

[1] Gulati A, Sharma B, Banerjee P, Mohan G. Getting More from Less: Story of India's Shrinking Water Resources. New Delhi: Indian Council for Research on International Economic Relations; 2019. p. 170. NABARD and ICRIER Report

[2] Burton MA, Sen R, Gordon-Walker S, Ghosh A. National Water Resources Framework Study: Roadmaps for Reforms. New Delhi: Council on Energy, Environment and Water; 2011

[3] Surendran U, Raja P, Jayakumar M, Subramoniam SR. Use of efficient water saving techniques for production of rice in India under climate change scenario: A critical review. Journal of Cleaner Production. 2021;**309**:127272

[4] Nayak AK, Tripathi R, Debnath M, Swain CK, Dhal B, Vijaykumar S, et al. Carbon and water footprint of rice, wheat & maize crop productions in India. Pedosphere. 2023;**33**(3):448-462. DOI: 10.1016/j.pedsph.2022.06.045

[5] Bhattacharyya R, Ghosh BN, Mishra PK, Mandal B, Rao CS, Sarkar D, et al. Soil degradation in India: Challenges and potential solutions. Sustainability. 2015;**7**(4):3528-3570

[6] Dobermann A. A critical assessment of the system of rice intensification (SRI). Agricultural Systems. 2004;**79**(3):261-281

[7] Vijayakumar S, Nayak AK, Ramaraj AP, Swain CK, Geethalakshmi V, Pazhanivelan S, et al. Rainfall and temperature projections and their impact assessment using CMIP5 models under different RCP scenarios for the eastern coastal region of India. Current Science. 2021;**121**(2):222

[8] Patil S, Reidsma P, Shah P, Purushothaman S, Wolf J. Comparing conventional and organic agriculture in Karnataka, India: Where and when can organic farming be sustainable? Land Use Policy. 2014;**37**:40-51

[9] Ambast SK, Tyagi NK, Raul SK. Management of declining groundwater in the trans Indo-Gangetic plain (India): Some options. Agricultural Water Management. 2006;**82**(3):279-296

[10] Singh O, Kasana A, Bhardwaj P. Understanding energy and groundwater irrigation nexus for sustainability over a highly irrigated ecosystem of north western India. Applied Water Science. 2022;**12**(3):44

[11] Anonymous. Indian Planning Commission. Eleventh Five-Year Plan: 2007-12. Volume-III. 2008, Agriculture, Rural Development, Industry, Services and Physical Infrastructure, 2008; pp. 1-537

[12] Mandavia AB. Modernization of Irrigation System Operational Management by Way of Canal Automation in India. Thailand: RAP Publication (FAO); 1999

[13] Madhok AK. Roorkee Enhancing Water Use Efficiency. Water Conclave 2020 Organized by Indian Institute of Technology Roorkee and National Institute of Hydrology, Roorkee during February 26-28. 2020;1-9

[14] Vijayakumar S, Kumar D, Ramesh K, Jinger D, Rajpoot SK. Effect of potassium fertilization on water productivity, irrigation water use efficiency, and grain quality under direct seeded rice-wheat cropping system. Journal of Plant Nutrition. 2022;**45**(13):2023-2038

[15] Anonymous. Creating an Ecosystem for Increasing Water-use Efficiency in Agriculture. New Delhi, India: Price Waterhouse Coopers Private Limited; 2021. pp. 1-51

[16] Anju A, Ravi SP, Bechan S. Water pollution with special reference to pesticide contamination in India. Journal of Water Resource and Protection. 2010;**2**(5):432-448

[17] Igboama WN, Hammed OS, Fatoba JO, Aroyehun MT, Ehiabhili JC. Review article on impact of groundwater contamination due to dumpsites using geophysical and physiochemical methods. Applied Water Science. 2022;**12**(6):130

[18] Nayak AK, Kumar A, Tripathi R, Panda BB, Mohanty S, Shahid M, et al. Improved Water Management Technologies for Rice Production System. NRRI Research Bulletin No. 32. Cuttack 753006, Odisha, India: ICAR-National Rice Research Institute; 2021. p. 40

[19] Hameed KA, Mosa AK, Jaber FA. Irrigation water reduction using system of Rice intensification compared with conventional cultivation methods in Iraq. Paddy and Water Environment. 2011;**9**(1):121-127

[20] Gathorne-Hardy A, Reddy DN, Venkatanarayana M, Harriss-White B. System of rice intensification provides environmental and economic gains but at the expense of social sustainability—A multidisciplinary analysis in India. Agricultural Systems. 2016;**143**:159-168

[21] Thakur AK, Mandal KG, Verma OP, Mohanty RK. Do system of rice intensification practices produce rice plants phenotypically and physiologically superior to conventional practice. Agronomy. 2023;**13**(4):1098

[22] Stoop WA, Uphoff N, Kassam A. A review of agricultural research issues raised by the system of rice intensification (SRI) from Madagascar: Opportunities for improving farming systems for resource-poor farmers. Agricultural Systems. 2002;**71**(3):249-274

[23] Tuong TP, Bouman BA. Rice production in water-scarce environments. In: Kijne JW, Barker R, Molden DJ, editors. Water Productivity in Agriculture: Limits and Opportunities for Improvement. Oxford, UK: CAB International; 2003. pp. 53-67

[24] Tabbal DF, Bouman BA, Bhuiyan SI, Sibayan EB, Sattar MA. On-farm strategies for reducing water input in irrigated rice; case studies in the Philippines. Agricultural Water Management. 2002;**56**(2):93-112

[25] Kima AS, Chung WG, Wang YM. Improving irrigated lowland rice water use efficiency under saturated soil culture for adoption in tropical climate conditions. Water. 2014;**6**(9):2830-2846

[26] Vijayakumar S, Karunakaran V, Gobinath R, Basavaraj K, Goud BR, Sha N, et al. Sustainable rice production in India through efficient water saving techniques. Chronicle of Bioresource Management. 2022;**6**(1):32-38

[27] Kima AS, Kima E, Bacyé B, Ouédraogo PA, Traore O, Traore S, et al. Evaluating supplementary water methodology with saturated soil irrigation for yield and water productivity improvement in semi-arid Rainfed Rice system, Burkina Faso. Sustainability. 2020;**12**(12):4819

[28] Yanti D, Berd I, Naspendra Z. The effect of water-saving irrigation on the growth of local rice plants. In: IOP Conference Series: Earth and Environmental Science. Indonesia: IOP Publishing; 2021;**922**(1):012036

[29] Vijayakumar S, Jinger D, Parthiban P, Lokesh S. Aerobic rice cultivation for enhanced water use efficiency. Indian Farming. 2018;*68*(6):3-6

[30] Prasad R. Aerobic rice systems. Advances in Agronomy. 2011;**111**:207-247

[31] Sahrawat KL. Organic matter accumulation in submerged soils. Advances in Agronomy. 2004;**81**:169-201

[32] Chivenge P, Angeles O, Hadi B, Acuin C, Connor M, Stuart A, et al. Ecosystem services in paddy rice systems. In: The Role of Ecosystem Services in Sustainable Food Systems. Elsevier Inc. Academic Press. 2020. pp. 181-201

[33] Awan MI, vanOort PA, Ahmad R, Bastiaans L, Meinke H. Farmers' views on the future prospects of aerobic rice culture in Pakistan. Land Use Policy. 2015;**42**:517-526

[34] Anapalli SS, Pinnamaneni SR, Reddy KN, Wagle P, Ashworth AJ. Eddy covariance assessment of alternate wetting and drying floodwater management on rice methane emissions. Heliyon. 2023;**9**(4):e14696

[35] Lampayan RM, Rejesus RM, Singleton GR, Bouman BA. Adoption and economics of alternate wetting and drying water management for irrigated lowland rice. Field Crops Research. 2015;**170**:95-108

[36] Subramanian E, Vijayakumar S, Sathishkumar A, Aathithyan C. Feasibility of drip irrigation in rice cultivation. Indian Farming. 2021;**71**(4):41-44

[37] Benouniche M, Kuper M, Hammani A, Boesveld H. Making the user visible: Analysing irrigation practices and farmers' logic to explain actual drip irrigation performance. Irrigation Science. 2014;**32**:405-420

[38] Subramanian E, Ramesh T, Vijayakumar S, Ravi V. Effect of drip irrigation on rice growth, yield attributes, grain yield and water use efficiency. The Indian Journal of Agricultural Sciences. 2023;**93**(4):371-375

[39] Sharda R, Mahajan G, Siag M, Singh A, Chauhan BS. Performance of drip-irrigated dry-seeded rice (*Oryza sativa* L.) in South Asia. Paddy and Water Environment. 2017;**15**:93-100

[40] Parthasarathi T, Vanitha K, Mohandass S, Vered E. Evaluation of drip irrigation system for water productivity and yield of rice. Agronomy Journal. 2018;**110**(6):2378-2389

[41] Chojnacka K, Witek-Krowiak A, Moustakas K, Skrzypczak D, Mikula K, Loizidou M. A transition from conventional irrigation to fertigation with reclaimed wastewater: Prospects and challenges. Renewable and Sustainable Energy Reviews. 2020;**130**:109959

[42] Sidhu HS, Jat ML, Singh Y, Sidhu RK, Gupta N, Singh P, et al. Sub-surface drip fertigation with conservation agriculture in a rice-wheat system: A breakthrough for addressing water and nitrogen use efficiency. Agricultural Water Management. 2019;**216**:273-283

[43] Ramachandrula VR, Kasa RR. Prevention and treatment of drip emitter clogging: A review of various innovative methods. Water Practice & Technology. 2022;**17**(10):2059-2070

[44] Fawibe OO, Honda K, Taguchi Y, Park S, Isoda A. Greenhouse gas emissions from rice field cultivation with drip irrigation and plastic film mulch.

Nutrient Cycling in Agroecosystems. 2019;**113**:51-62

[45] Belaud G, Mateos L, Aliod R, Buisson MC, Faci E, Gendre S, et al. Irrigation and energy: Issues and challenges. Irrigation and Drainage. 2020;**69**:177-185

[46] Pramanik M, Khanna M, Singh M, Singh DK, Sudhishri S, Bhatia A, et al. Automation of soil moisture sensor-based basin irrigation system. Smart Agricultural Technology. 2022;**2**:100032

[47] Champness M, Ballester-Lurbe C, Filev-Maia R, Hornbuckle J. Smart sensing and automated irrigation for sustainable rice systems: A state of the art review. Advances in Agronomy. 2023;**177**:259-285

[48] Lee J. Evaluation of automatic irrigation system for Rice cultivation and sustainable agriculture water management. Sustainability. 2022;**14**(17):11044

[49] Barkunan SR, Bhanumathi V, Sethuram J. Smart sensor for automatic drip irrigation system for paddy cultivation. Computers & Electrical Engineering. 2019;**73**:180-193

[50] Eshete DG, Sinshaw BG, Legese KG. Critical review on improving irrigation water use efficiency: Advances, challenges, and opportunities in the Ethiopia context. Water-Energy Nexus. 2020;**3**:143-154

[51] Bazaluk O, Havrysh V, Nitsenko V, Mazur Y, Lavrenko S. Low-cost smart farm irrigation systems in Kherson province: Feasibility study. Agronomy. 2022;**12**(5):1013

[52] Trinovita ZJ, Nurtam MR, Laksmana I, Putera P, Putri IA. The implementation of the internet of things (IoT) on an automatic plant watering and fertilizing system based on solar electricity. In: IOP Conference Series: Earth and Environmental Science. Malang, Indonesia: IOP Publishing; 2022;**1097**(1):012057

[53] Sales N, Remédios O, Arsenio A. Wireless sensor and actuator system for smart irrigation on the cloud. In: 2015 IEEE 2nd World Forum on Internet of Things (WF-IoT). United State: IEEE; 2015. pp. 693

[54] Ferrarezi RS, Peng TW. Smart system for automated irrigation using internet of things devices. HortTechnology. 2021;**31**(6):642-649

[55] Vijayakumar S, Choudhary AK, Deiveegan M, Thirumalaikumar R, Kumar RM. Android based mobile application for rice crop management. Chronicle of Bioresource Management. 2022;**6**(1):19-24

[56] Saravanane P, Pavithra M, Vijayakumar S. Weed management in direct seeded rice. Indian Farming. 2021;*71*(04):61-64

[57] Ishfaq M, Akbar N, Anjum SA, Anwar-Ijl-Haq M. Growth, yield and water productivity of dry direct seeded rice and transplanted aromatic rice under different irrigation management regimes. Journal of Integrative Agriculture. 2020;**19**(11):2656-2673

[58] Bista B. Direct seeded rice: A new technology for enhanced resource-use efficiency. International Journal of Applied Sciences and Biotechnology. 2018;**6**(3):181-198

[59] Mahender A, Anandan A, Pradhan SK. Early seedling vigour, an imperative trait for direct-seeded rice: An overview on physio-morphological parameters and molecular markers. Planta. 2015;**241**:1027-1050

[60] Keerthi DE, Saravanane P, Poonguzhalan R, Nadaradjan S, Muthukumarasamy S, Vijayakumar S. Effect of brown manuring practices on yield, nutrient dynamics and soil micro-flora in wet seeded rice in the coastal deltaic ecosystem. Oryza—An International Journal of Rice. 2022;**59**(4):519-524

[61] Vijayakumar S, Gobinath R, Jasudas Gompa S, Surekha K, Kumar RM, Sundaram RM. Management of micro-nutrient deficiency and toxicity in rice. Indian Farming. 2022;**72**(10):11-14

[62] Pooja K, Saravanane P, Sridevi V, Nadaradjan S, Vijayakumar S. Effect of cultivars and weed management practices on productivity, profitability and energetics of dry direct-seeded rice. Oryza—An International Journal of Rice. 2021;**58**(3):442-447

[63] Kumar V, Singh AK, Jat SL, Parihar CM, Pooniya V, Sharma S, et al. Influence of site-specific nutrient management on growth and yield of maize (*Zea mays*) under conservation tillage. Indian Journal of Agronomy. 2014;**59**(4):657-660

[64] Vijayakumar S, Kumar D, Jinger D, Bussa B, Panda BB. 4R nutrient stewardship based potassium. Indian Farming. 2021;**71**(01):22-25

[65] Subramanian E, Aathithyan C, Raghavendran VB, Vijayakumar S. Optimization of nitrogen fertilization for aerobic rice (*Oryza sativa*). Indian Journal of Agronomy. 2020;**65**(2):180-184

[66] Jinger D, Dass A, Kumar V, Kaur R, Kumari K. Weed management strategies in the climate change era. Indian Farming. 2016;**66**(9):09-13

[67] Ramesh K, Kumar SV, Upadhyay PK, Chauhan BS. Revisiting the concept of the critical period of weed control. The Journal of Agricultural Science. 2021;**159**(9-10):636-642

[68] Saha S, Munda S, Satapathy BS, Vijayakumar S, Poonam A. Weed Management Technology for Rice. Cuttack, Odisha: National Rice Research Institute; 2021

[69] Akhter J, Monneveux P, Sabir SA, Ashraf MY, Lateef Z, Serraj R. Selection of drought tolerant and high water use efficient rice cultivars through 13C isotope discrimination technique. Pakistan Journal of Botany. 2010;**42**(6):3887-3897

Chapter 8

Application of Geospatial Techniques in Agricultural Resource Management

Syeda Mishal Zahra, Muhammad Adnan Shahid, Zahid Maqbool, Rehan Mahmood Sabir, Muhammad Safdar, Muhammad Danish Majeed and Aneela Sarwar

Abstract

Although technological advancements have sparked the beginning of the fourth agricultural revolution, human beings are still facing severe problems such as shrinking croplands, dwindling water supplies, negative consequences of climate change, and so on in achieving agricultural resilience to meet the demands of the growing population over the globe. Geospatial techniques involving the integrated use of geographic information system (GIS), remote sensing (RS), and artificial intelligence (AI) provide a strong basis for sustainable management of agricultural resources aimed at increased agricultural production. In recent times, these advanced tools have been increasingly used in agricultural production at local, regional, and global levels. This chapter focuses on the widespread application of geospatial techniques for agricultural resource management by monitoring crop growth and yield forecasting, crop disease and pest infestation, land use and land cover mapping, flood monitoring, and water resource management. Moreover, we also discuss various methodologies involved in monitoring and mapping abovementioned agricultural resources. This chapter will provide deep insight into the available literature on the use of geospatial techniques in the monitoring and management of agricultural resources. Moreover, it will be helpful for scientists to develop integrated methodologies focused on exploring satellite data for sustainable management of agricultural resources.

Keywords: geospatial techniques, agricultural resource management, crop growth monitoring, flood monitoring, land cover mapping

1. Introduction

Agriculture is the backbone of the economy, particularly in developing countries, and plays a pivotal role in economic stability and ensuring food security. As a matter of fact, the global population is increasing exponentially and is expected to reach 10 billion by 2050. This drastic increase in global population puts pressure on agricultural resources for more food production despite the constraints of the environment [1, 2].

Changes in climatic conditions, restrictions on expanding agricultural land, and scarcity of water are among the major complications for increasing crop production. Therefore, cropland management, protection of soil quality, judicious use of water, and increase in species diversity are the few basic steps for sustainable management of agricultural resources to meet future food demands [3]. More importantly, monitoring of crop growth and health coupled with timely intervention strategies are necessary for enhanced agricultural crop production along with a reduction in the use of input resources in accomplishing global food security goals.

Technological advances in recent years have shown much development in sensor technology, communication systems, and data analysis tools. Such advances in technologies have facilitated the sustainable use of agricultural input variables along with mitigation of future losses while promoting increased and sustainable yield to accomplish global food security goals. Numerous existing and newly developed techniques and tools such as Geographic Information System (GIS), Artificial Intelligence (AI), Remote Sensing (RS), Global Navigation Satellite Systems (GNSS), Internet of Things (IoT), and Big Data Analytics (BDA) have become essential for global food security through effective analysis of crops as well as soils. When merged with other sources of evidence, these technology solutions are supplying data-driven information and insight for focused or site-specific management of crops, trying to ensure higher production [4]. GIS is a type of database system that includes tools for gathering, storage, and information retrieval in addition to tools for analyzing, converting, and demonstrating spatiotemporal data for a particular intention [5–7]. GIS is one of the key technologies that gives the spatial frame of reference and information on numerous attributes, each of which can be accessed as a separate data layer. Furthermore, it supplies the tools necessary to modify both spatial and non-spatial data and illustrates the results in map formats that are both intuitive and demonstrative [8]. GIS has widespread applications in environmental studies, medical services, and resource management in food, industrial, and agricultural sectors. In recent times, GIS tools have been increasingly used in agricultural production at local, regional, and global levels. GIS together with other techniques such as RS, GNSS, and data analytics has been widely utilized to facilitate crop and soil interventions [9–11].

Remote sensing is a type of geospatial technique that involves acquisition of information about an object or phenomenon without making physical contact with the object using space-borne or airborne satellites. Remote sensing technique captures images of the earth's surface in different wavelength regions of Electro-Magnetic Spectrum (EMS). Some of the images symbolize reflected solar radiation in visible and near infrared regions; a thermal infrared wavelength region represents estimations of energy emitted by the earth's surface, and a microwave region represents a measure of relative return from the earth's surface. The basic mechanism of remote sensing technology involves emitted electromagnetic radiations coming from sun, falling on different objects on earth and reflected to atmosphere, which is captured by the distantly placed satellite sensor as images of different wavelengths. However, the process involved in acquiring information about an object or area of interest by means of measuring radiant energy reflected or emitted by an object or surface has been summarized in **Figure 1**. The complete processing of remote sensing involves various steps including (1) emission of electromagnetic radiations; (2) energy transmission from source to earth's surface; (3) interactions of EMR with the earth's surface, that is, reflection and emission; (4) energy transmission from the surface to satellite sensors; (5) sensor data output; and (6) transmission, procession, and analysis of satellite data. This remotely sensed data is comprised of spatial information (size, shape, and

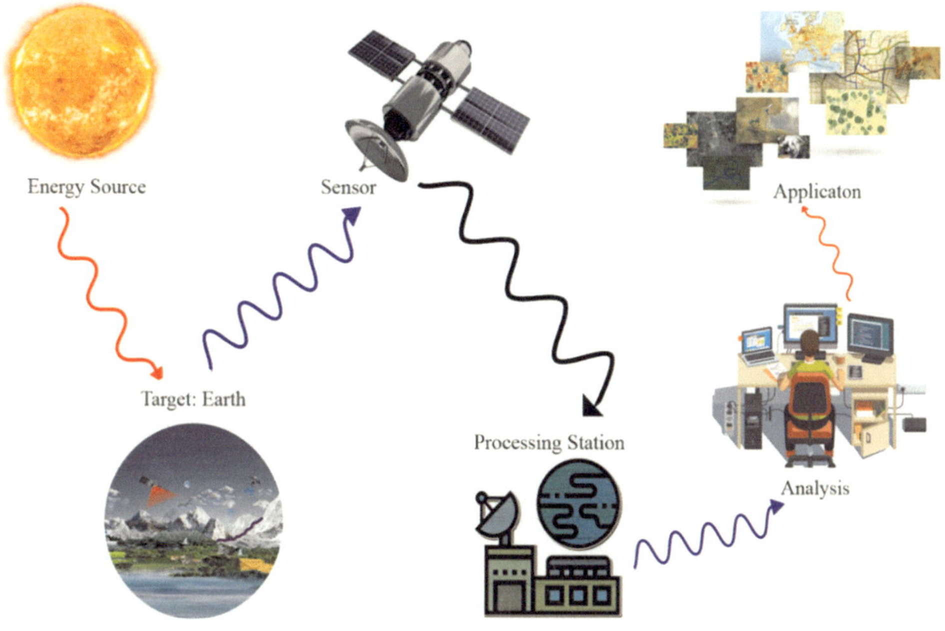

Figure 1.
Concept of remote sensing and GIS - applications of remote sensing.

orientation) as well as spectral information (tone, color, and spectral signature). As it is known, GIS creates, manages, analyzes, and maps all types of data. GIS connects data to a map, integrating location data (where things are) with all types of descriptive information (what things are like there). This provides a basis for mapping and analysis in almost every industry and assists users to understand patterns, relationships, and geographic context. The benefits include improved communication and efficiency as well as better management and decision making. Although RS and GIS have been widely utilized in various domains of agriculture such as soil mapping, terrain analysis, crop stress mapping, crop yield mapping, soil nutrient assessment, and organic matter mapping, it has been suggested that the potential of high-resolution remotely sensed data to deal with spatiotemporal variations make it valuable to be utilized in precision agriculture. When it comes to precision agriculture, various combinations of spatial resolutions, spectral coverage, and frequency ranges are utilized. It has been reported that monitoring of crop growth and mapping of crop yield can be achieved using low-resolution satellite data; however, mapping the severity of disease infestation necessarily requires fine-resolution satellite data. Literature suggests that RS and GIS are the core of precision agriculture because they collect, store, retrieve, and analyze feature- and location-based information and provide statistics alternatives particularly for site-specific monitoring [12].

Since its inception in the 1960s, remote sensing technology has been widely used in agriculture [13]. Several global and national agricultural monitoring systems based on remote sensing platforms are now in operation. These systems deliver consistent, timely, and important information to agricultural producers, managers, and policymakers. Crop identification and cropland mapping; crop growth monitoring and yield estimation/prediction; inversion of key biophysical, biochemical, and environmental parameters; crop damage/disaster monitoring; precision farming; and so

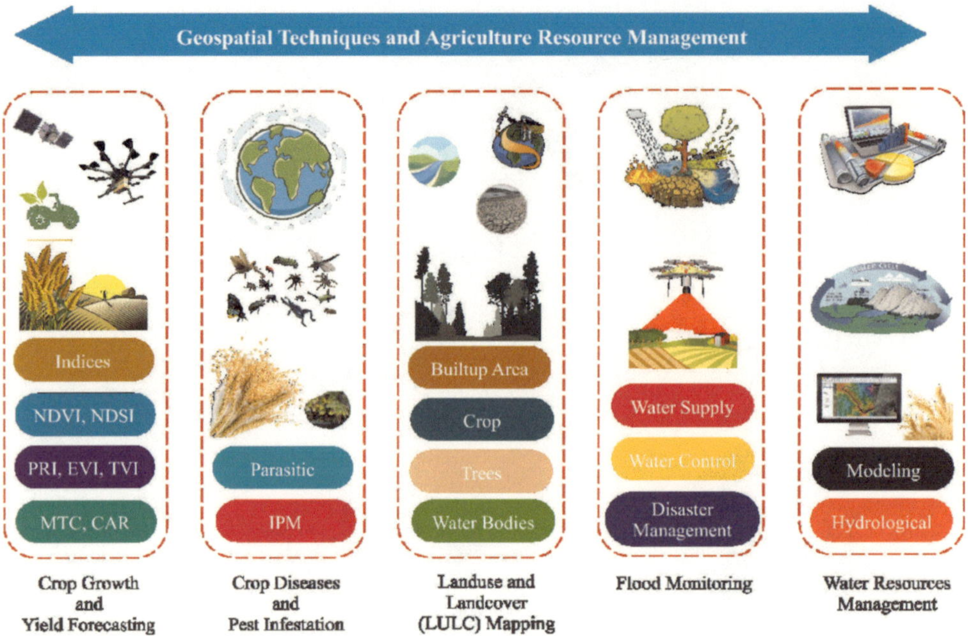

Figure 2.
Different indices/ parameters for monitoring and management of various agricultural resources.

on are all important applications of geospatial techniques in agriculture resource monitoring. This chapter is focused on the use of geospatial techniques for agricultural resource management. The various indices and other parameters determined using geospatial techniques for mapping and monitoring of crop growth and yield forecasting, crop disease and pest infestation, nutrient deficiency, land use, land cover mapping, flood monitoring, and water resource management are summarized in **Figure 2**.

Digital maps based on RS data and GIS differ broadly from traditional maps. In these maps, satellite data is utilized to map a given attribute such as soil analysis, crop yield, precipitation, nutrient availability, pest infestation, and so on. Satellite-based positioning or navigation system provides information about longitude, latitude, and altitude. GIS provides sophisticated abilities by using statistical techniques and geospatial predictive analysis to retrieve inter-attribute interactions. Such observations are beneficial for decision making and management of resources on a sustainable basis. These digital maps help farmers and researchers to identify hotspot regions, map field boundaries and water bodies, and recognize the relationship of different features inside and outside the field's boundary lines. High-fidelity field visualization allows site-specific implementation of nutrient content, herbicides, pesticides; irrigation to improve productivity; and to reduce input costs. In this chapter, we present the scientific work associated with the use of remote sensing and GIS techniques for sustainable management of agricultural resources. This chapter is organized in such a way that it starts with highlighting the need of RS and GIS technologies for mapping and monitoring agricultural resources. Afterward, research presents the recent developments in RS and GIS technologies involved in agricultural resource management.

2. Crop growth and yield forecasting

Monitoring development of crops and making estimates of their yields are both extremely vital for any nation to be able to make decisions on food production for its consistently expanding population. Typically, crop yield forecasting across crop growth includes models that integrate global climatic and soil conditions and other ancillary data as executed accordingly to characterize the development, photosynthetic activity, evapotranspiration, and yield of a particular crop relying on reliable physical and physiological theories. However, these models are a poor significant predictor when variations in soils, stress, pressure, or management practices are present. The process of anticipating crop growth and initial crop yield over agricultural fields is a crucial method for both the planning of food security and the prediction of agricultural economic return. Monitoring the growth of crops and estimating their output have been easier thanks to the ongoing development of RS and GIS technologies, which have led to improvements in both the process and procedures involved [14–16]. Estimation of crop yield has been demonstrated by multiple studies to benefit from the integration of GIS and RS technology. The research carried out by Memon et al. [17] evidenced how efficient integrating hyperspectral Landsat satellite imagery and contrasting various RS-based spectroscopy indices was in monitoring crop growth and early crop yield forecasting over agricultural fields. The focus of this research was on the impact of rice yields. The information can be useful in the long-term planning of agricultural sustainability in cropping systems involving rice and wheat. The findings of the study that was conducted by Hassan and Goheer [18] demonstrated that an accurate early forecast of wheat crop yield prior to growing crops could be achieved through the utilization of vegetation indices deduced from modest resolution imaging spectroradiometer satellite imagery in conjunction with crop yield information and a strategy that is based on GIS modeling. Muslim et al. [19] employed a GIS-based environmental policy-integrated climate model in yet another research. This model served as an effective tool for forecasting rice yield and was based on integrated climate policy. Several of the most popular vegetation indices (VIs) used in remote sensing are included in **Table 1** along with details such as their mathematical formulations, the scales they are designed for, and the factors they assess. These vegetative indices have several applications, including crop monitoring, stress detection, biomass estimation, and yield forecasting. However, the correct index must be selected considering the given task and surrounding factors.

The prototype regional- and local-crop-level data, soil data, agricultural production data, and climatic data to spatial and temporal approximate variability in crop yield. In a similar manner, Al-Gaadi et al. [16] predicted potato tuber crop yield by extracting the normalized difference vegetation indices and the soil-adjusted vegetation index from Landsat satellite data obtained during the different phases of potato growth. Simulations of crop yield prediction that are premised on GIS and RS have the potential to have broad applicability in the process of notifying agricultural production regulations that are geographically centered. For instance, based on the results of these modeling techniques, compliance review can be designed to change the contributing factors to crop yield and quality (that include farm management strategies, wind patterns, availability of water, altitude, terrain, plant health, and inclusive growth) [20, 21]. This is possible because policy intervention is a form of intervention. It is essential to make yield projections for crops well in advance of harvest time, especially in areas where there is a high degree of climatic variability. Planning and

Index	Scale	Parameter	Formula
Normalized Difference Vegetation Index (NDVI) [14, 15, 19–21]	Canopy	Biomass; Vegetation Fraction	$\frac{NIR-Red}{NIR+Red}$
NDVI is used for assessing the robustness and density of vegetation. It has important applications in crop growth monitoring, land-use planning, and ecological research. NDVI values closer to 1 suggest that the plants are in better health [22].			
Green Normalized Difference Vegetation Index (GNDVI) [23–26]	Canopy	Chlorophyll; Vegetation Fraction	$\frac{NIR-Green}{NIR+Green}$
GNDVI is a tool developed based on satellite data, frequently used for assessment of crop health and growth. Higher GNDVI values indicate more robust plant life. Values between 0 and 1 are possible [27].			
Photochemical Reflectance Index (PRI) [27–32]	Canopy	Photosynthesis efficiency/ RUE	$\frac{R570-R531}{R570+R531}$
PRI is a valuable tool for assessing photosynthetic activity and stress in plants. PRI reflects changes in pigment composition of leaves and the quantity of light that is absorbed by chlorophyll. PRI values are directly proportional to vegetation health and its numerical scale ranges from −1 to 1, where higher values indicates better health [33].			
Normalized Difference Red Edge (NDRE) [34–37]	Canopy	Chlorophyll/ Nitrogen	$\frac{R790-R720}{R790+R720}$
NDRE is a vegetation index developed using geospatial data for assessing plant health and biomass production. NDRE index determines the alterations in vegetation cover, encompassing factors such as plant growth, stress, and nutrient insufficiencies [27].			
Canopy Chlorophyll Content Index (CCCI) [38–42]	Canopy	N Status/ Chlorophyll	$\frac{NDRE-NDRE_{min}}{NDRE_{max}-NDRE_{min}}$
CCCI is widely employed for estimating the chlorophyll concentration in plant canopies. CCCI exhibits sensitivity toward variations in leaf chlorophyll concentrations and structural changes in leaves. CCCI values ranges from 0 to 1, where higher values are indicative of higher levels of chlorophyll content [43].			
Ratio Vegetation Index (RVI) [44–47]	Leaf	Biomass	$\frac{NIR}{Red}$
RVI is widely used to assess the health and growth of vegetation for the purpose of evaluating crop productivity and identifying nutrient insufficiencies. RVI values ranges from 0 to 1 where higher VHI values are indicator of healthier vegetation [48].			
Enhanced Vegetation Index (EVI) [49–52]	Canopy/ Regional	Biomass/Vegetation Cover	$2.5 * \frac{NIR-Red}{NIR+C1*Red-C2*Blue+L}$ C1 = 6; C2 = 7.5; L = 1
EVI is a vegetation index that enables the assessment of vegetation density, cover, and health and frequently utilized in land-use and ecological studies. EVI values ranges from −1 to 1 where elevated values are indicator of more robust and healthier vegetation [53].			
Visible Atmospherically Resistant Index (VARIgreen) [53, 54]	Canopy/ Regional	Vegetation Fraction/ LAI	$\frac{Green-Red}{Green+Red+Blue}$
VARIgreen is utilized to assess vegetation cover and health and primarily used in regions where atmospheric interference is of significant concern. VARIgreen index values ranges from −1 to 1, with higher values indicating a greater degree of vegetation health [48].			

Table 1.
Shot description of frequently used vegetation indices (VIs).

policymaking for food security and the prediction of agricultural economic return should include monitoring the conditions under which agricultural crops grow as well as making projections about the potential crop yield [14, 15, 19]. This could include the development of policies for increasing the efficiency of agricultural production and maintaining its sustainable development [15]. To provide food for a growing population in low- and middle-income countries (LMICs), agricultural production systems need to work toward closing the yield gap that exists between the yields that farmers currently accomplish and the yields that may be conceivably easily achievable in a rainfed basic sustenance agricultural system. In order to evaluate and map crop yield gaps, Hochman et al. [21] created a simulation that consolidated numerical yield and crop varieties area data, remotely sensed data, cropping information processing paradigm, and GIS mapping. The purpose of this model was to address the discrepancy between the two sets of data.

There exist multiple approaches for predicting crop yield; some of the approaches are given in **Figure 3**. The conventional approach to predicting crop yield involves the assessment of crop condition by professionals. Throughout the crop-growing season, various observations and measurements are conducted, including but not limited to tiller number, spikelet number and fertility percentage, pest and fungal damage percentage, and weed infestation percentage, among others. The data obtained through this method can be utilized to make predictions about yield through the application of regression techniques or by drawing upon the insights of local experts [22]. Two additional techniques employed for predicting crop production are remote sensing and crop simulation models. The aim of the yield forecast is to provide accurate, scientifically sound, and impartial predictions of crop yield at the earliest possible stage in the growing season, considering the impact of weather and climate. The

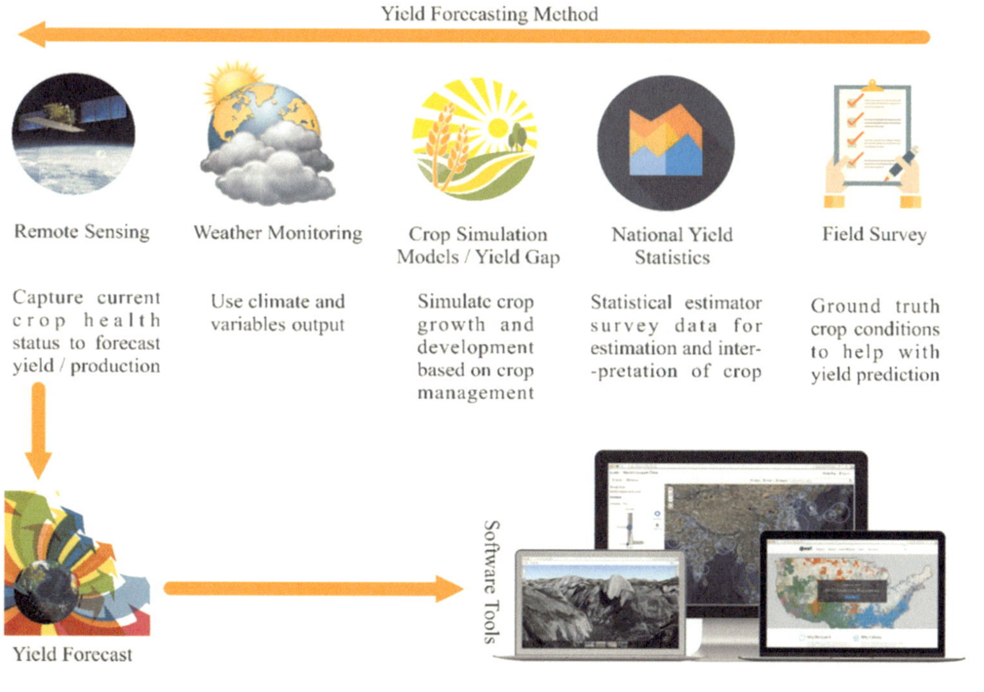

Figure 3.
Approaches used for yield forecasting include field survey, national yield statistics, crop simulation, weather monitoring, and remote sensing.

disparities between forecasts and final estimates lie in the temporal aspect of their dissemination. Predictions are generated prior to the complete harvest of the crop, whereas evaluations are conducted after the harvest of the crop. The outcomes are derived from the implementation of a statistical estimator on the survey data, and the subsequent point estimates are analyzed by commodity.

Throughout history, farmers have consistently engaged in the practice of making predictions as a means of strategizing their agricultural methods. The selection of a cultivar, the optimal planting timeframe, and the appropriate quantity of fertilizer to utilize are contingent upon the prevailing climatic conditions. If farmers are aware of a high probability of rainfall in the upcoming week, they will promptly proceed to sow their seeds in the field [23, 24]. The accurate prediction of crop yield necessitates the concurrent prediction or knowledge of other significant parameters. An instance of this would be the measurement of the land that was sown at the commencement of the cultivation period and the measurement of the land that was reaped. The utilization of geospatial methodologies in the prediction of crop development and yield has numerous advantages, such as the capacity to furnish farmers, policymakers, and other interested parties with precise and prompt information. The utilization of this information can aid in the process of decision making, specifically in tasks such as identifying the most favorable periods for planting, choosing suitable crop types, and optimizing the management of resources, such as water and fertilizer, for greater efficiency.

3. Crop disease and pest infestation

It is essential that there should be no instance of hunger or malnutrition among the population if the nation is to experience long-term economic expansion and environmental sustainability. Malnutrition is the only cause of one-third of deaths and underweight, with retarded growth among children less than 5 years of age in poor nations. Food insecurity, poor maternal and childcare practices, a lack of access to safe drinking water, and insufficient quality sanitation and health services are all contributors to malnutrition, which is caused by a deficiency in both macronutrients and micronutrients, as well as an imbalance in nutritional intake and disease [25]. Over the past few centuries, the mismatch between the increasing food demand of the world's booming population and the amount of agricultural supply has made the issue worse [26]. The main limiting variables that can influence agricultural productivity, growth, and eventually food security are both biotic and abiotic constraints. These include unfavorable and harsh temperatures, a lack of water, bad soil, insect infestation, illnesses, and weeds. In addition, changes in temperature or climatic disruption have a significant and deleterious effect on agricultural output and productivity around the globe, particularly in tropical regions [27]. As the world's climate continues to shift, experts predict that agricultural output will drop by 20 percent by 2080 and that the situation would be even direr in developing nations like India. Droughts, floods, and other weather extremes are just a few examples of how climate change is affecting crop production. Insect pests and disease epidemics have become more common because of these shifts and disruptions [28]. Exports and imports of agricultural products have increased worldwide since the liberalization that followed the founding of the World Trade Organization (WTO) in 1995. The spread of invading alien species, which threaten native and introduced plant communities worldwide, has been accelerated by the globalization of trade. As a result of global warming and trade liberalization, more infectious diseases and insect pest outbreaks have appeared. Furthermore, new tools have made it possible to reduce the impact on agricultural yield as much as possible [28]. Now, more than ever, GIS and remote sensing are being used extensively for precise data collection and analysis when it comes to crops as shown in **Figure 4**. In this section, it will look at how geospatial technology has played an admirable part in the detection and mitigation of plant/crop diseases and insect pests. This section may also illustrate the supplementary understanding of how geospatial technologies would be utilized in agriculture to lessen the impact of the disease and insect-pest attack.

The term "healthy" is used to describe a plant that is fully realizing its genetic potential in terms of physiological effects. When insects or diseases cause havoc on plant life, normal physiological processes are thrown off [29]. The plant's health, biomass, and output are all negatively impacted by pests and illnesses carried by insects. Typically, the effects of insect pests and diseases on yield and quality are not immediately apparent throughout the growing season. Forest diseases and pests can have a massive effect on ecological services, altering natural landscaping and its cultural value, hindering wildlife habitat and biodiversity, and limiting the forest's

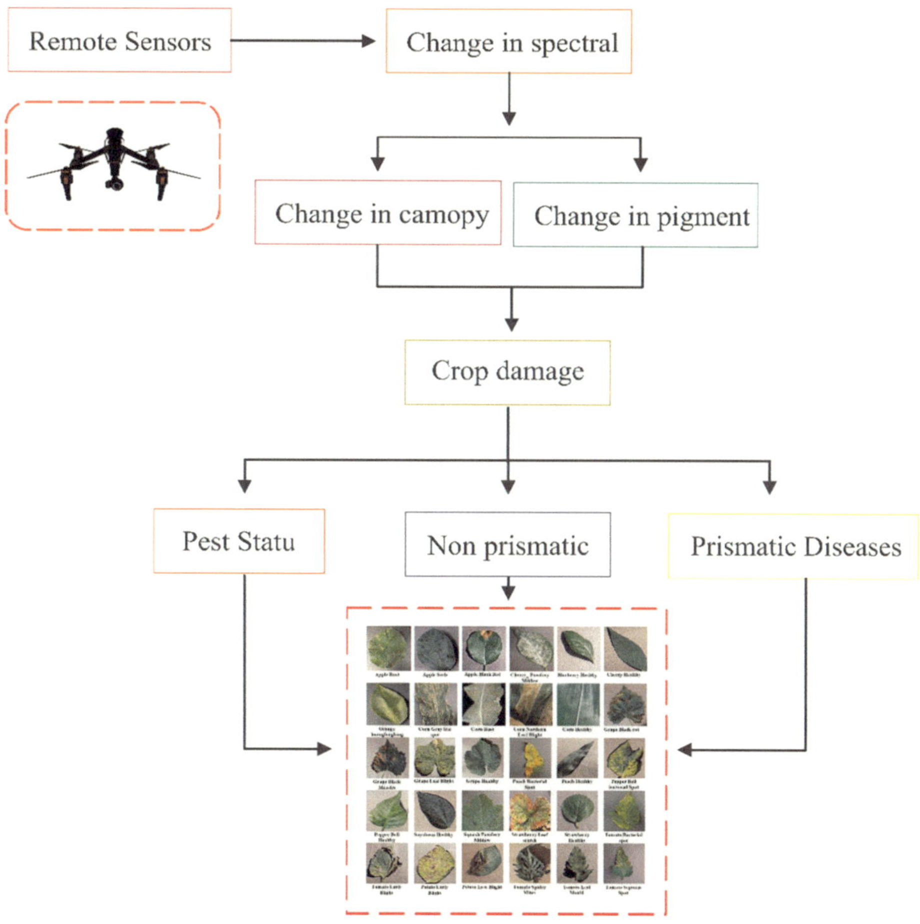

Figure 4.
The emergence of crop-damaging diseases and pests, and the use of remote sensing.

capacity for carbon sequestration, all of which harm agricultural output. GIS stores, captures, analyzes, and displays data that describes an area of the earth's surface. Most common GIS features include data storage and retrieval, presentation and querying, evaluation (geometric and thematically), conversion, and others. Information gathering, processing, administration, and display are the four pillars upon which a GIS is built [30]. The proliferation of data-capture methods is proportional to the variety of technological tools at our disposal. As a result of advancements in geospatial techniques like GPS, geodetic surveys, laser scanners, aerial photography, and satellite-based remote sensing, and so on, it is now much simpler to keep tabs on plant health. Arizona, Northern California, Idaho, Washington, and Oregon used GIS/GPS technology to analyze semi-automated data [31].

Using a unified strategy, we geo-referenced weather data, plant growth data, and satellite imagery in a GIS to assess the potential for pests and diseases as well as the optimal growing conditions for our crops. Using elevation, weather, and satellite data projections, projections for both favorable growing conditions and the spread of illness were developed for agricultural regions. Models for six insect pests and twelve crop diseases were computed and displayed daily in geo-referenced maps for agricultural survey regions. Dates and yields of grape harvests were also forecasted with remarkable precision. Information was disseminated daily via the Internet as regional weather, insect, and disease risk maps based on data derived from GIS [30]. Difficulties associated with plant nourishment, insect identification, crop forecasting, water requirements, and weed management are just some of the many areas where remote sensing has been utilized. Changes in plant characteristics are indicators of disease and pest attack, which can be detected by remote sensing. Typical ground-based optical scouting techniques may miss tiny variations in vegetation, but a remote sensor's capacity to do so makes it a helpful tool for assessing within-field variability, evaluating crop development, and managing fields based on present conditions [30].

Nowadays, precision agriculture relies on geospatial technologies like GPS integration with GIS for Insect-Pest Management (IPM). Crop management planning benefits further from the utilization of remote sensing and spatial analysis, as well as other technologies such as the Global Navigation Satellite System (GNSS) [32]. However, there are also some drawbacks to these techniques. For example, high spatial resolution imaging is not easily accessible for all locations, especially rural ones. While hyper-spectral photography is essential, consultancies and end-users/farmers often lack specialist knowledge of geospatial techniques, and timely mapping and near-real-time picture collection and delivery processes are crucial to the success of this field of study. Despite these difficulties, geospatial technology can also provide useful data in an IPM setting by facilitating a thorough comprehension of the relative spatial of the abiotic and biotic characteristics of a field and its crop, as well as giving information on the populations of diseases and pests that are either already present or likely to appear [30]. In a field where reports of pest and disease outbreaks on a massive scale are on the rise, this technology is proving invaluable. Improved production, pest and disease control, and other facets of forestry management are all possible thanks to geospatial technologies. These tools are now used mostly in academic settings, but with improved access to data and technology, this is likely to change. Increased yields and lessened concerns over food insecurity are two of the many benefits that have resulted from the widespread adoption of geospatial data and technology.

Crop output must rise along with population growth to guarantee that everyone, no matter where they live, has access to adequate food and water. Agricultural sustainability is achieved via the employment of a variety of technologies and strategies that make production safer and reduce negative effects on the environment. Successful agriculture relies heavily on the use of precision agriculture tools. Using data gathered by geospatial technologies, precision farming is a method of micromanagement that improves agricultural and territorial management decisions. Pests and diseases have a significant impact on crop yields, which in turn affects the ability to feed the world's population. But the development of geospatial technology has facilitated the operations of controlling many pests and illnesses affecting plant health.

4. Land use/land cover (LU/LC) mapping

The terms "land cover" and "land use" are sometimes used indiscriminately; nevertheless, the true connotations of these concepts are significantly different from one another. The covering surface on the earth is referred to as land cover, and it might include things like flora, urban development, waterways, fallow land, or other things. To effectively carry out global monitoring investigations, resource planning endeavors, and development endeavors, it is essential to recognize, delineate, and map land cover [8, 33, 34]. The establishment of a benchmark from which monitored operations, such as change detection, can be carried out and the provision of ground cover data for baseline thematic layers are both accomplished through the process of identifying land cover [35]. However, the term "land use" relates to the use that the land is put to, such as for agricultural purposes, for the goal of providing an environment for wildlife, or for other recreational activities. Because it is necessary to have up-to-date data to determine what percentage of land is now being put to what kind of use and what kinds of changes occur in land utilization throughout the year; approaches for land use need both mappings as a baseline and ongoing monitoring [33, 36, 37]. The dynamic characterization of the world's LU/LC has been summarized in **Table 2**. Having this knowledge aids in the process of formulating plans to strike a balance among preservation efforts, competing uses, and the demands of development. Land use studies are being conducted because of several issues including the loss of forested land, the reduction in the amount of fertile land, and creeping urbanization [11].

Resource planning and control, preservation of wildlife habitats, baseline modeling for GIS inputs, urban expansion and encroaching, route and logistical planning for

LULC - Objects	Characteristics
Water bodies	Possesses very minimal to almost no sparse vegetation, no rock outcrops, and no man-made structures such as docks [44]
Trees	Any considerable concentration of thick vegetation is usually characterized by a covered or thick canopy [35]
Grass	Uniform regions of grass-like flora, sometimes known as blade-type leaves, provide the impression of being distinct from tree cover and sparse vegetation. Grasslands that grow naturally in the environment without any visible human planning [36]
Waterlogged vegetation	Regions with any kind of flora where water and soil are visibly mixed. Regions that are flooded periodically consist of a combination of grass, shrubs, trees, and barren land [55]
Crops	Cereal grains, grasslands, and crops were all seeded and plotted by humans [19]

LULC - Objects	Characteristics
Shrubs and scrub	Landscaping with uncovered rock and soil should have either isolated pockets of vegetation or isolated plants distributed across the terrain. Obstacles that prevent that are populated by shrubs among dense forests that are obviously not higher than the trees. Look grayer and browner because of a less dense layer of leaves [34]
Built-up area	Commercial, business, and private buildings along with the parking areas constructed by humans that are enclosed together with a variety of housing developments, including structures, roads, gardens, and trees, as well as domestic constructions or housing developments that are separated and bordered by land covered with vegetation [48]
Fallow land	Stretches of rock or dirt that are covered in very little or no flora at all having vast tracts of land consist primarily of sand and desert, with little or no wildlife [56]
Snow and ice	Large, evenly distributed expanses of snow or ice are often only seen in mountainous regions or at the highest latitude and evenly distributed across a wide area [57]

Table 2.
Dynamic categorization of the world's land use and land cover (LULC).

geophysical investigation, resource exploitation operations, disaster characterization (storms, inundation, volcano, earthquake, and fires), legislative borders for taxation and asset appraisal, object tracking (identifying landing fields, highways, hillsides, and crossings), and delineating of the land/water boundary are the important tasks that can be performed using LU/LC data gathered using geospatial techniques. However, in agricultural resource management, LU/LC refers to the physical characteristics and use of land in a particular area, including vegetation cover, soil type, water availability, and land use activities as shown in **Figure 5**.

Understanding LU/LC is crucial for sustainable agricultural management, as it provides information on land use change, crop suitability, and environmental impacts. In agriculture resource management, LU/LC data is used in various ways, including:

Crop Suitability Analysis: LU/LC data can be used to identify areas suitable for specific crops. For example, if a farmer is interested in planting a particular crop, they can use LU/LC data to identify areas with suitable soil types, water availability, and

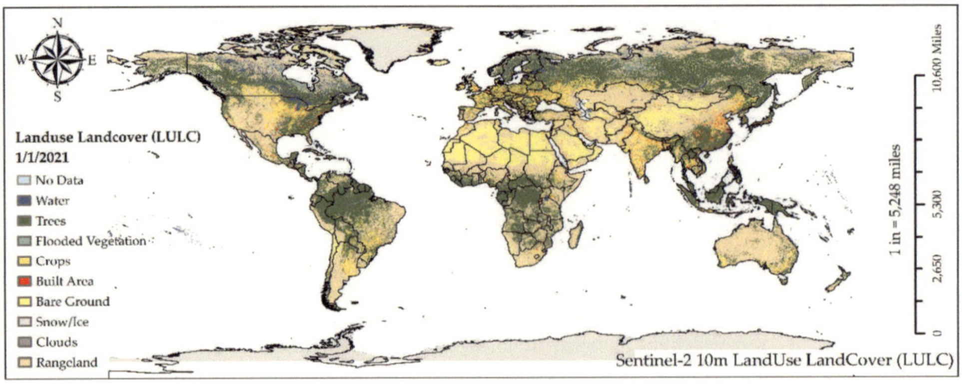

Figure 5.
Global 10 m land use/ land cover (LU/LC) map for 2021 using geospatial technologies [source: ESRI website (https://www.esri.com/partners/impact-observatory-a2T5x000008 4pJXEAY/land-use-land-cover–a2d 5x000005juReAAI)].

vegetation cover for that crop. This analysis helps to optimize crop production, leading to higher yields and better resource management.

Land Use Planning and Management: LU/LC data are also used in land use planning and management. By understanding the current and past use of land, farmers and land managers can develop land use plans that maximize productivity while minimizing environmental impacts. For example, if a particular area has a history of intensive agriculture, land managers may decide to rotate crops to avoid soil degradation and nutrient depletion.

Environmental Monitoring: LU/LC data are also essential for environmental monitoring. By tracking changes in land use and vegetation cover, environmental impacts such as deforestation, soil erosion, and habitat loss can be identified and addressed. This information is crucial for developing effective conservation strategies and ensuring sustainable resource management.

Natural Resource Management: LU/LC data are also used to manage natural resources such as water and soil. By understanding the characteristics of the land, farmers can implement practices such as soil conservation, water harvesting, and irrigation that optimize resource use and minimize waste.

Disaster Management: LU/LC data are also essential for disaster management. By identifying areas at risk of natural disasters such as floods, landslides, and droughts, farmers and land managers can implement measures to mitigate the impact of these events. For example, farmers may decide to plant crops that are more resistant to drought or implement soil conservation practices to reduce soil erosion during floods. So, LU/LC data are essential in agriculture resource management and provide crucial information on crop suitability, land use planning and management, environmental monitoring, natural resource management, and disaster management. By using LU/LC data, farmers and land managers can optimize resource use, minimize environmental impacts, and ensure sustainable agricultural practices. The efficacy of remote sensing (RS) data in facilitating land use and land cover (LU/LC) mapping has been substantiated by scientific literature. The use of RS data has been shown to provide precise and current data on land use changes, thereby aiding sustainable land management practices.

5. Flood monitoring

Floods are one of the most destructive environmental disasters worldwide, contributing to more substantial economic and social consequences compared to any other natural event. Because of their widespread distribution, floods cause more death and destruction than any other natural calamities [38]. When it rains heavily in a place, downstream areas frequently become inundated because of the excess water. Whenever water flows overland and into low-lying areas, this can cause flooding [39]. Floods and other environmental disasters wreak havoc on both human and natural resources and generate significant amounts of destruction [40, 41]. Flooding has an adverse impact on the lives of around 140 million citizens each year on average. Assessment of extensive flooding is of the utmost significance regarding the repercussions on society, the economy, and the environment [42]. Flood control and mitigation efforts are essential to minimize the damage that can be caused to environmental assets, agricultural land, and other infrastructures, among other things [43, 44]. As a result, doing a study of the likelihood that an area will be flooded is a crucial responsibility for advance warning systems and emergency responders as they work toward developing management plans for preventing and mitigating future flood disasters [45]. For instance, HAZUS is a GIS-based natural hazard assessment tool formulated for evaluating flood hazards; HEC-FDA is a computer program to facilitate agricultural engineers across vulnerability assessment of flooding-risk-reduction strategy. These are just two examples of the many comprehensive techniques that are currently available and utilized by many data analysis organizations around the world. GIS is a computer-based system that handles georeferenced data by providing the capability for inputs, information management (data management and access), modification and evaluation, and dissemination. GIS offers a wide variety of techniques that may be used to determine the region that has been impacted by floods and to make predictions on the areas that are expected to be inundated because of high-water levels in rivers [41]. GIS will be utilized to a significant degree to compile material obtained from a variety of maps, aerial pictures, satellite images, and digital elevation models (DEM) [44]. The vulnerability assessment also will appropriate utilization of the data from the census as well as any other appropriate statistical abstracts to better cater to the requirements of the native population. The research and technique of collecting data (spectrum, geographical, and temporally) about a region or material thing without making direct physical touch with the entity or region under inquiry are known as remote sensing [42–44].

In recent years, remote sensing technology and geographic information systems have emerged as the two most important tools for flood inundation mapping [43]. The demarcation of flood zones and the creation of flood danger and flood risk mapping for places that are prone to flooding are the two primary focuses of attention in this sector of industry [42]. Any endeavor at flood evaluation or hazard modeling in this region is hampered by the limited availability of high-resolution DEMs, even though river flooding in developing countries is very severe because of their overreliance on agriculture. However, the accessibility of high-resolution DEMs is limited. In places prone to flooding, flood hazard mapping is an integral part of carrying out the necessary planning for land use [45]. It generates charts and maps that are simple to understand and can be accessed quickly, which makes it easier for administrators and planners to identify areas of risk and prioritize their activities to mitigate or respond to such risks [46]. Worldwide natural disasters due to flood hazard, the hotspot areas, flood frequency, distribution and deciles using geospatial technologies are illustrated in **Figure 6.**

The traditional methodologies of predicting floods were not trustworthy enough to make a precise forecast. The techniques used for flood monitoring are illustrated in **Figure** 7. But usage of remote sensing and GIS technologies is extremely helpful in determining and evaluating flood-related consequences caused by excessive rainfall in a catchment area or by ocean wave surges in coastal areas. During the previous year, numerous researchers and engineers put various models to use in order to estimate flood levels [46]. At the present time, geospatial approaches offer a diverse selection of data sources that can be utilized for the modeling of floods. In the field of disaster modeling, including flood inundation mapping, cartography, and the analysis of complicated problems, the analytic hierarchy process, abbreviated as AHP, is among the most well-known and successful methodologies [40, 43]. In addition to the Analytic Hierarchy Process (AHP), other recognized and appropriate models for hazard assessment include the Multi-Criteria Decision Support Approach (MCDA), the Weights of Evidence (WoE), the logistic regression (LR) model, the responsive neuro-fuzzy interface system, artificial neural networks (ANN), and the FR model [47, 48]. Researchers attempted to demonstrate a connection between the predictor variables and the

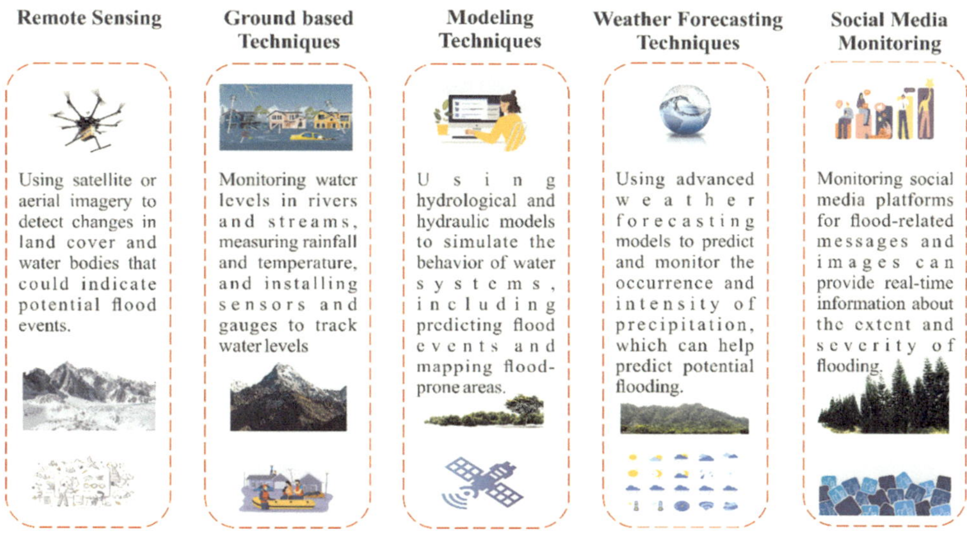

Figure 6.
Global natural disaster hotspots for flood hazard frequency and distribution using geospatial technologies.

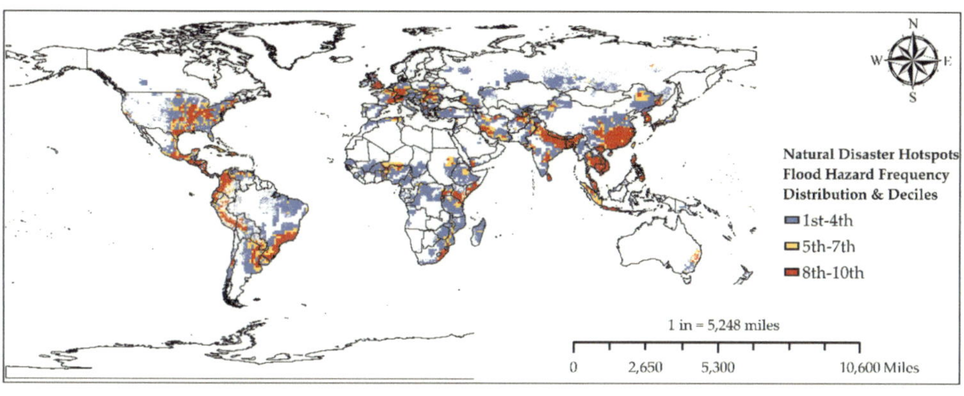

Figure 7.
Geospatial techniques and global map for flood hazard [58] [online source: SEDAC website (https://sedac.ciesin.columbia.edu/data/set/ndh-flood-hazard-frequency-distribution/maps)].

occurrence of the flood by using the ANN approach, which was used in the forecasting of flood [48]. It has been stated that the ANN approach can deal with all inputs that are unpredictable to derive useful information. The employment of MCDA, RS, and GIS techniques is particularly helpful in the study and mapping of possible flood-prone regions because these techniques are dependable and precise [48]. The MCDA method is applicable for flood assessment and mapping in areas with insufficient or no available data, and it may be of use to local authorities in the process of flood mitigation. In China, the AHP model was used for the purpose of flood deflection [47, 48].

The modeling of floods is a complicated process that requires considering a great number of different parameters. The application of the RS approach makes a substantial contribution to the mapping and evaluation of flood risks. Because they are rapid and more effective, RS and GIS offer the greatest opportunity to catch, preserve, integrate, alter, access, analyze, and present information for the assessment of prospective risk regions. This is especially useful in situations where time is of the essence. This study is an example of an ensemble method that demonstrates how effective GIS-based flood modeling can be. The frequency ratio (FR) method, GIS, and remote sensing were all utilized in the process of estimating the likelihood of flooding.

6. Water resource management

Water is a fundamental element of nature's resources and can be found in many different varieties, including underground aquifers, rivers, streams, lakes, glaciers, and so on. The capability of water to keep humans alive, transport nutrition to agricultural production, neutralize pollutants and hazardous substances, and regulate the hydrological regime are the primary reasons for the significance of water [46]. Consequently, the management of water resources is an important problem for us to address now to lessen the likelihood of water shortages for coming generations. Substantial oscillations in the hydrological system are currently causing degradation or water contamination supplies, as illustrated in **Table 3**. These high variations are a direct result of the significant economic and demographic shifts that have occurred in recent times [45]. As a direct consequence of this, this invaluable asset is currently subject to competition and requires both protection and management for conservation. In this section, we have presented the critical issue of gaining knowledge of water resources through the application of geo-informatics by utilizing different types of intelligence, including classic, enhanced, composite, and artificial [49].

Water Resources Management	Hydrological management
	Modeling of groundwater
	Quality analysis of water
	Water supply management
	Sewer system management
	Storm water control
	Flood disaster management

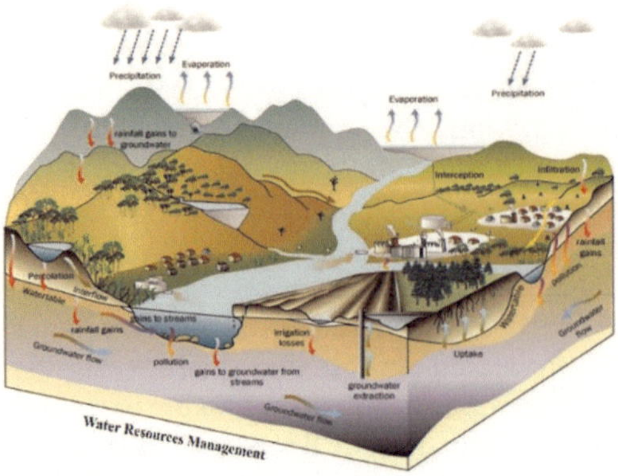

Table 3.
Water resources management strategies.

The primary purpose of water resource management is to comprehend and manage water resources, whereas geospatial approaches make use of remote sensing, geographic information systems, and global positioning system to make this process more efficient. In addition to its function as a factor in maintaining life, water also plays a part in the generation of potentially lethal risks in the manner of floods and droughts [11, 43, 48]. GIS databases document and catalog water supply information. There are servers located all around the world where the data accumulated on water resources are kept. Some of the details are generally the product of GIS data acquisition [11]. So, with the aid of GIS, researchers can store and distribute massive amounts of information about water resources. Large geospatial satellites launched from outside the Earth's atmosphere are interconnected with GIS and utilized to aid in the global dissemination of data/information. When a base station needs to obtain geographical data, the satellite makes it available to them wirelessly. Fortunately, cloud-based deployment options are available for most GIS systems today. This allows information stored in any GIS server to be accessed by geospatial centers located anywhere in the world. The widespread availability and adaptability of data and information access are features of GIS applications [50].

Research published on water has revealed that water alters both its condition and pressure throughout the course of time [1, 45, 51]. When it is necessary to keep monitoring various water circumstances, GIS becomes an extremely useful tool. Therefore, hydrologists are among the professionals who stand to benefit the most from geographic information systems. Using a GIS that has been carefully constructed, one can do a variety of studies on water. For instance, hydrogeology is a field of study that analyzes groundwater along with its preservation, incidence, and mobility properties. The nature and properties of water that is either stored deep inside the earth or that is on the surface and is either motionless or moving can be entered into GIS as data, stored there, and retrieved for use in the system's subsequent processing. This can be done for water that is either underground or on the surface. The process of modeling groundwater entails hydrologists trying to comprehend the behavior and features of groundwater. Keeping in mind the limited supply of water, there is a lot of research that can be done to safeguard water catchment zones [52]. In addition to this, GIS can assist in the production of models and designs that can assist in the responsible utilization of underground water. When conducting investigations and case studies, the application of geospatial techniques can allow the creation of digital photographs of groundwater [53]. On this planet, there are many kinds of water, and not all of them are fit for human or animal consumption. Studies on aspects such as slope, drainage features, and patterns of land exploitation can all be analyzed with GIS to determine whether the water in a certain region is suitable for consumption by humans. Sample data can be analyzed, saved, and used to report generation thanks to the capacity of GIS to manage vast volumes of data sets. GIS also has the capability to handle the generation of maps [54]. These studies can be utilized by the relevant organization, or even by the authorities, to make future studies and laws on water, as well as to assess whether the water is suitable for human consumption.

7. Conclusion

The present chapter aims at highlighting the importance and usage of geospatial techniques (remote sensing together with GIS technologies) in agricultural resource management, including monitoring and mapping of crop growth and production forecasting, crop disease and pest infestation, land use and land cover mapping, flood monitoring, and water resource management. Literature suggests that the use of remote sensing (RS) and geographic information systems (GIS) in agriculture has been increasing rapidly over the past few decades due to advancement of digital technologies that have leveraged GIS as a vital alliance technology for evaluating crops, soils, and their environments. In this chapter, we discussed various analyses and methodologies involved in using remotely sensed satellite data for management of agriculture resources. Based on the scientific literature, it can be concluded that geospatial techniques have been effectively utilized for digital mapping of agricultural resources and play a crucial role in ensuring sustainable agricultural production. It has been observed that multispectral satellite sensors are preferably used in mapping and monitoring of agricultural resources because of their low cost but have limited application, which is linked with their low spatial and spectral resolution. However, hyperspectral satellite sensors with fine spatial and spectral resolution allow more promising results with greater accuracy and provide exceptional abilities to address enormous challenges associated with regular monitoring and mapping of agricultural resources at large areas. This book chapter will be helpful to develop cost-effective and accurate monitoring solutions for agriculture resource management at fine spatial resolution for decision making especially in land use issues in agriculture crop production.

Acknowledgements

We acknowledge the funding provided by Higher Education Commission (HEC) Pakistan for the establishment of Agricultural Remote Sensing Lab at University of Agriculture, Faisalabad. We also acknowledge the support of Vice Chancellor, UAF, for providing us with infrastructure facilities at University of Agriculture, Faisalabad.

Conflict of interest

All the authors and co-authors declare that there is no conflict of interest.

Author details

Syeda Mishal Zahra[1,2,3], Muhammad Adnan Shahid[1,2], Zahid Maqbool[1*], Rehan Mahmood Sabir[1,2], Muhammad Safdar[1,2], Muhammad Danish Majeed[1,2] and Aneela Sarwar[4]

1 Agricultural Remote Sensing Lab (ARSL), National Center for GIS and Space Applications, University of Agriculture, Faisalabad, Pakistan

2 Faculty of Agricultural Engineering and Technology, Department of Irrigation and Drainage, University of Agriculture, Faisalabad, Pakistan

3 International Water Management Institute (IWMI), Lahore, Pakistan

4 Department of Agricultural Engineering, Bahauddin Zakariya University, Multan, Pakistan

*Address all correspondence to: zahid.maqbool@uaf.edu.pk

References

[1] Ghosh P, Kumpatla SP. GIS applications in agriculture. In: Zhang Y, Cheng Q, editors. Geographic Information Systems and Applications in Coastal Studies. London, UK: IntechOpen; 2012. pp. 1-27

[2] FAO. The Future of Food and Agriculture: Trends and Challenges. Rome: FAO Publishing; 2017;**4**:1951-1960

[3] Gomiero T. Soil degradation, land scarcity and food security: Reviewing a complex challenge. Sustainability. 2016; **8**(3):1-41

[4] Delgado JA, Short NM, Roberts DP, Vandenberg B. Big data analysis for sustainable agriculture on a geospatial cloud framework. Frontiers in Sustainable Food Systems. 2019;**3**:54-59

[5] Delgado JA, Vandenberg B, Neer D, D'Adamo R. Emerging nutrient management databases and networks of networks will have broad applicability in future machine learning and artificial intelligence applications in soil and water conservation. Journal of Soil and Water Conservation. 2019;**74**(6): 113A-118A

[6] Gomiero T. Alternative land management strategies and their impact on soil conservation. Agriculture. 2013; **3**(3):464-483

[7] Koroleva EV, Nikitin YY. U-max-statistics and limit theorems for perimeters and areas of random polygons. Journal of Multivariate Analysis. 2014;**127**:98-111

[8] Mathenge M, Sonneveld BGJS, Broerse JEW. Application of GIS in agriculture in promoting evidence-informed decision making for improving agriculture: A systematic review. Sustainability. 2022;**14**(16):9974-9999

[9] Kingra PK, Majumder D, Singh SP. Application of remote sensing and GIS in agriculture and natural resource management under changing climatic conditions. Agricultural Research Journal. 2016;**53**(3):295

[10] Kazemi H, Akinci H. A land use suitability model for rainfed farming by multi-criteria decision-making analysis (MCDA) and geographic information system (GIS). Ecological Engineering. 2018;**116**:1-6

[11] Gebeyehu MN. Remote sensing and GIS application in agriculture and natural resource management. International Journal of Environmental Sciences & Natural Resources. 2019; **19**(2):45-49

[12] Kumar S, Karaliya SK, Chaudhary S. Precision farming technologies towards enhancing productivity and sustainability of Rice-wheat cropping system. International Journal of Current Microbiology and Applied Sciences. 2017;**6**(3):142-151

[13] Sishodia RP, Ray RL, Singh SK. Applications of remote sensing in precision agriculture: A review. Remote Sensing. 2020;**12**:3136-3167

[14] Houshyar E, Smith P, Mahmoodi-Eshkaftaki M, Azadi H. Sustainability of wheat production in Southwest Iran: A fuzzy-GIS based evaluation by ANFIS. Cogent Food & Agriculture. 2017;**3**(1): 1327682. DOI: 10.1080/23311932.2017.1327682

[15] Mann ML, Warner JM. Ethiopian wheat yield and yield gap estimation: A

spatially explicit small area integrated data approach. Field Crops Research. 2017;**201**:60-74

[16] Al-Gaadi KA, Hassaballa AA, Tola E, Kayad AG, Madugundu R, Alblewi B, et al. Prediction of potato crop yield using precision agriculture techniques. PLoS One. 2016;**11**(9):1-16

[17] Memon MS, Jun Z, Sun C, Jiang C, Xu W, Hu Q, et al. Assessment of wheat straw cover and yield performance in a rice-wheat cropping system by using Landsat satellite data. Sustainability. 2019;**11**(19):5369-5390

[18] Hassan SS, Goheer MA. Modeling and monitoring wheat crop yield using geospatial techniques: A case study of Potohar region, Pakistan. Journal of the Indian Society of Remote Sensing. 2021; **49**(6):1331-1342

[19] Muslim M, Romshoo SA, Rather AQ. Paddy crop yield estimation in Kashmir Himalayan rice bowl using remote sensing and simulation model. Environmental Monitoring and Assessment. 2015;**187**(6):1-12

[20] Wang M, Tao FL, Shi WJ. Corn yield forecasting in Northeast China using remotely sensed spectral indices and crop phenology metrics. Journal of Integrative Agriculture. 2014;**13**(7): 1538-1545

[21] Hochman Z, Gobbett D, Holzworth D, McClelland T, van Rees H, Marinoni O, et al. Reprint of quantifying yield gaps in rainfed cropping systems: A case study of wheat in Australia. Field Crops Research. 2013;**143**:65-75

[22] Basso B, Cammarano D, Carfagna E. Review of crop yield forecasting methods and early warning systems. In: Proceedings of the First Meeting of the Scientific Advisory Committee of the Global Strategy to Improve Agricultural and Rural Statistics. July 2013; FAO Headquarters, Rome, Italy: Scientific Research Publishing; 2013. p. 19

[23] USDA. The Yield Forecasting Program of NASS the Statistical Methods Branch. Washington, D.C.: Statistics Division, National Agricultural Statistics Service, U.S. Department of Agriculture; 2012. p. 104

[24] Lobell DB, Burke MB. On the use of statistical models to predict crop yield responses to climate change. Agricultural and Forest Meteorology. 2010;**150**(11): 1443-1452

[25] Tirado MC, Crahay P, Mahy L, Zanev C, Neira M, Msangi S, et al. Climate change and nutrition: Creating a climate for nutrition security. Food and Nutrition Bulletin. 2013;**34**(4):533-547

[26] Savary S, Ficke A, Aubertot JN, Hollier C. Crop losses due to diseases and their implications for global food production losses and food security. Food Security. 2012;**4**(4):519-537

[27] Säre M. Climate change and agricultural production. Culture and Environment. 2019;**4**:201-208

[28] Masters G, Norgrove L. Climate change and invasive alien species. CABI Working Paper. 2010;**1**:1-30

[29] Sabtu NM, Idris NH, Ishak MHI. The role of geospatial in plant pests and diseases: An overview. In IOP Conference Series: Earth and Environmental Science. 2018;**169**(1):012013

[30] Anwer A, Singh G. Geo-spatial technology for plant disease and insect pest management. Bulletin of Environment, Pharmacology and Life Sciences. 2019;**8**:1-12

[31] Thomas CS, Skinner PW, Fox AD, Greer CA, Gubler WD. Utilization of GIS/GPS-based information technology in commercial crop decision making in California, Washington, Oregon, Idaho, and Arizona. Journal of Nematology. 2002;**34**(3):200-206

[32] Qu Y, Tao B. The constitution of vegetable traceability system in agricultural IOT. Journal of Chemical and Pharmaceutical Research. 2014;**6**(7): 2580-2583

[33] Mishra PK, Rai A, Rai SC. Land use and land cover change detection using geospatial techniques in the Sikkim Himalaya, India. Egyptian Journal of Remote Sensing and Space Science. 2020;**23**(2):133-143

[34] Souza CM, Shimbo JZ, Rosa MR, Parente LL, Alencar AA, Rudorff BFT, et al. Reconstructing three decades of land use and land cover changes in brazilian biomes with Landsat archive and earth engine. Remote Sensing. 2020; **12**(17):2735-2762

[35] Naikoo MW, Rihan M, Shahfahad PAH, Talukdar S, Mallick J, Ishtiaq M, et al. Analysis of peri-urban land use/land cover change and its drivers using geospatial techniques and geographically weighted regression. Environmental Science and Pollution Research. 2022;**29**:1-9. DOI: 10.1007/s11356-022-18853-4

[36] Jin S, Yang L, Zhu Z, Homer C. A land cover change detection and classification protocol for updating Alaska NLCD 2001 to 2011. Remote Sensing of Environment. 2017;**195**:44-55

[37] Phiri D, Morgenroth J. Developments in Landsat land cover classification methods: A review. Remote Sensing. 2017;**9**(9):1-10

[38] Wood HM. The use of earth observing satellites for hazard support. International Geoscience and Remote Sensing Symposium (IGARSS). 2001;**1**: 135-137

[39] Merz B, Kreibich H, Schwarze R, Thieken A. Review article "assessment of economic flood damage." Natural Hazards and Earth System Science. 2010; **10**(8):1697-1724

[40] Du J, Fang J, Xu W, Shi P. Analysis of dry/wet conditions using the standardized precipitation index and its potential usefulness for drought/flood monitoring in Hunan Province, China. Stochastic Environmental Research and Risk Assessment. 2013;**27**(2):377-387

[41] Youssef AM, Pradhan B, Hassan AM. Flash flood risk estimation along the St. Katherine road, southern Sinai, Egypt using GIS based morphometry and satellite imagery. Environmental Earth Sciences. 2011;**62**(3):611-623

[42] Markantonis V, Meyer V, Lienhoop N. Evaluation of the environmental impacts of extreme floods in the Evros River basin using contingent valuation method. Natural Hazards. 2013;**69**(3):1535-1549

[43] Billa L, Shattri M, Mahmud AR, Ghazali AH. Comprehensive planning and the role of SDSS in flood disaster management in Malaysia. Disaster Prevention and Management: An International Journal. 2006;**15**(2): 233-240

[44] Huang X, Tan H, Zhou J, Yang T, Benjamin A, Wen SW, et al. Flood hazard in Hunan province of China: An economic loss analysis. Natural Hazards. 2008;**47**(1):65-73

[45] Tehrany MS, Pradhan B, Jebur MN. Flood susceptibility mapping using a

novel ensemble weights-of-evidence and support vector machine models in GIS. Journal of Hydrology. 2014;**512**:332-343

[46] Kisi O, Nia AM, Gosheh MG, Tajabadi MRJ, Ahmadi A. Intermittent streamflow forecasting by using several data driven techniques. Water Resources Management. 2012;**26**(2):457-474

[47] Setiawati MD, Miura F, Aryastana P. Validation of Hourly GSMAP and Ground Base Estimate of Precipitation for Flood Monitoring in Kumamoto, Japan. In: Geospatial Technology for Water Resource Applications. Boca Raton: CRC Press Taylor & Francis Group; 2016. pp. 130-143

[48] Ahansal Y, Bouziani M, Yaagoubi R, Sebari I, Sebari K, Kenny L. Towards smart irrigation: A literature review on the use of geospatial technologies and machine learning in the Management of Water Resources in arboriculture. Agronomy. 2022;**12**(2):297

[49] Thakur PK, Aggarwal SP, Nikam BR, Garg V, Chouksey A, Dhote PR. Training, education, research and capacity building needs and future requirements in applications of geospatial technology for water resources management. International Archives of the Photogrammetry, Remote Sensing and Spatial Information Sciences - ISPRS Archives. 2018;**42**(5): 29-36

[50] Hathout S. The use of GIS for monitoring and predicting urban growth in east and west St Paul, Winnipeg, Manitoba. Canada. Journal of Environmental Management. 2002; **66**(3):229-238

[51] Wang HB, Wu SR, Shi JS, Li B. Qualitative hazard and risk assessment of landslides: A practical framework for a case study in China. Natural Hazards. 2013;**69**(3):1281-1294

[52] Ingole NA, Ram RN, Ranjan R, Shankhwar AK. Advance application of geospatial technology for fisheries perspective in Tarai region of Himalayan state of Uttarakhand. Sustainable Water Resources Management. 2015;**1**(2): 181-187

[53] Jayakrishnan R, Srinivasan R, Santhi C, Arnold JG. Advances in the application of the SWAT model for water resources management. Hydrological Processes: An International Journal. 2005;**19**(3):749-762

[54] Pourghasemi HR, Moradi HR, Fatemi Aghda SM, Gokceoglu C, Pradhan B. GIS-based landslide susceptibility mapping with probabilistic likelihood ratio and spatial multi-criteria evaluation models (north of Tehran, Iran). Arabian Journal of Geosciences. 2014;**7**(5):1857-1878

[55] Moisa MB, Dejene IN, Merga BB, Gemeda DO. Impacts of land use/land cover dynamics on land surface temperature using geospatial techniques in Anger River sub-basin. Western Ethiopia. Environmental Earth Sciences. 2022;**81**(3):1-14

[56] Chen J, Chen J, Liao A, Cao X, Chen L, Chen X, et al. Global land cover mapping at 30 m resolution: A POK-based operational approach. ISPRS Journal of Photogrammetry and Remote Sensing. 2015;**103**:7-27

[57] James E. A land use and land cover classification system for use with remote sensor data. Remote Sensing. 2001;**2**:1-12

[58] Dilley M, Chen RS, Deichmann U, Lerner-Lam AL, Arnold M. Natural Disaster Hotspots: A Global Risk Analysis. Washington, DC: World Bank Publications; 2005